William Miles
**Numerical Methods with Python**

# Also of Interest

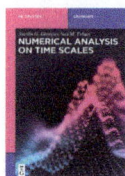

*Numerical Analysis on Time Scales*
Svetlin G. Georgiev and Inci M. Erhan, 2022
ISBN 978-3-11-078725-2, e-ISBN (PDF) 978-3-11-078732-0

*Advanced Mathematics*
*An Invitation in Preparation for Graduate School*
Patrick Guidotti, 2022
ISBN 978-3-11-078085-7, e-ISBN (PDF) 978-3-11-078092-5

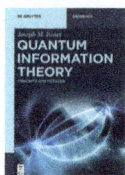

*Quantum Information Theory*
*Concepts and Methods*
Joseph M. Renes, 2022
ISBN 978-3-11-057024-3, e-ISBN (PDF) 978-3-11-057025-0

*Multi-level Mixed-Integer Optimization*
*Parametric Programming Approach*
Styliani Avraamidou, Efstratios Pistikopoulos, 2022
ISBN 978-3-11-076030-9, e-ISBN (PDF) 978-3-11-076031-6

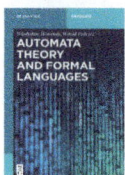

*Automata Theory and Formal Languages*
Wladyslaw Homenda and Witold Pedrycz, 2022
ISBN 978-3-11-075227-4, e-ISBN (PDF) 978-3-11-075230-4

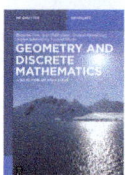

*Geometry and Discrete Mathematics*
*A Selection of Highlights*
Benjamin Fine, Anja Moldenhauer, Gerhard Rosenberger,
Annika Schürenber, Leonard Wienke, 2022
ISBN 978-3-11-074077-6, e-ISBN (PDF) 978-3-11-074078-3

William Miles

# Numerical Methods with Python

for the Sciences

DE GRUYTER

**Mathematics Subject Classification 2010**
Primary: 34-04, 35-04; Secondary: 92C45, 92D25, 34C28, 37D45

**Author**
William Miles, PhD
Stetson University
Department of Mathematics and Computer Science
421 N. Woodland Blvd.
Deland 32723
FL USA
wmiles@stetson.edu

ISBN 978-3-11-077645-4
e-ISBN (PDF) 978-3-11-077664-5
e-ISBN (EPUB) 978-3-11-077693-5

**Library of Congress Control Number: 2022950763**

**Bibliographic information published by the Deutsche Nationalbibliothek**
The Deutsche Nationalbibliothek lists this publication in the Deutsche Nationalbibliografie;
detailed bibliographic data are available on the Internet at http://dnb.dnb.de.

This book is dedicated to Andi and Emmy. To have them in my life is a blessing beyond measure.

# Acknowledgment

I thank William Wood and Sammi Smith for taking the time to read and edit large sections of the text. I also thank my students, Halle Block, Emily Mehigan, and Casey Ramey, for alerting me of errors as they took the course.

https://doi.org/10.1515/9783110776645-201

# Contents

Acknowledgment —— VII

1        Introduction —— 1

2        The basic operations in Python —— 3
2.1        Obtaining Python —— 3
2.2        Addition, subtraction, multiplication, and division —— 3
2.3        Powers —— 8
2.4        Displaying output —— 8
2.5        Exercises —— 10

3        Functions —— 11
3.1        Exponentials, logs, and trig functions —— 11
3.2        Variables —— 13
3.3        Defining and using mathematical functions —— 17
3.4        Getting input from the keyboard —— 19
3.5        Graphing functions —— 20
3.6        Exercises —— 38

4        Matrices, vectors, and linear systems —— 41
4.1        Matrices with numpy —— 41
4.1.1        Addition and subtraction: $A \pm B$ —— 44
4.1.2        Component-wise multiplication: $A * B$ —— 46
4.1.3        Component-wise division: $A/B$ —— 47
4.1.4        Scalar multiplication: $cA$ —— 48
4.1.5        Standard matrix multiplication —— 49
4.2        Matrix inversion —— 51
4.2.1        The identity matrix —— 52
4.2.2        The inverse of a matrix —— 53
4.3        Linear systems —— 57
4.4        Exercises —— 68

5        Iteration —— 71
5.1        Finding roots: the bisection method —— 71
5.2        Euler's method for differential equations —— 80
5.2.1        Systems of differential equations and higher-order differential
           equations —— 91
5.2.2        Interpolation—using the approximations —— 99
5.3        Exercises —— 101

**6**      **Statistics —— 103**
6.1      File handling —— **103**
6.2      Descriptive statistics —— **118**
6.3      Probability —— **123**
6.3.1      Numerical integration —— **125**
6.4      Confidence interval for the mean of a population —— **132**
6.5      Hypothesis testing —— **140**
6.6      Comparing groups —— **147**
6.6.1      Comparing means of two groups —— **147**
6.6.2      Comparing means of more than two groups —— **152**
6.7      Exercises —— **157**

**7**      **Regression —— 161**
7.1      Linear regression —— **161**
7.1.1      Correlation —— **170**
7.1.2      Multiple linear regression —— **171**
7.2      Logistic regression —— **178**
7.2.1      Digit recognition model —— **186**
7.3      Neural networks —— **193**
7.4      Exercises —— **206**

**A**      **Python code —— 209**
A.1      Chapter 2 code —— **209**
A.2      Chapter 3 code —— **209**
A.3      Chapter 4 code —— **216**
A.4      Chapter 5 code —— **221**
A.5      Chapter 6 code —— **228**
A.6      Chapter 7 code —— **240**

**B**      **Solutions —— 249**

**Index —— 313**

**Index of Python Commands —— 315**

# 1 Introduction

After years of mentoring undergraduate student research projects, it is clear that the most popular projects are applied in nature. It is also true that most "real-world" problems can not be solved explicitly. That is, we cannot find a nice, neat formula to solve the problem. Because of this, we must use numerical techniques to determine a close approximation to the solution of the problem of interest. These techniques often require us to repeat a process hundreds or thousands of times in order for the approximation to be "close enough" to the actual solution or for the approximation to evolve for the desired length of time. In addition to such repeated processes, we also frequently need to handle large amounts of data or manipulate large matrices in order to arrive at a solution. To solve the types of problems that arise in math and science, we frequently need to develop and implement an *algorithm*. An algorithm is the definition of a process that is to be used in solving a problem. Generally, algorithms are presented as a list of steps to be followed in order to arrive at a solution. In this book, we introduce some of the fundamental ideas and methods that are used to solve scientific problems. Some of the most frequently occurring challenges include:

– the need to locate the extreme values of a function;
– the need to solve large linear systems;
– the need to solve differential equations (or systems of differential equations);
– the need to draw conclusions about a population based on a sample (inferential statistics);
– the need to find the "best" linear model for a set of data (linear regression); and
– the need to classify objects (logistic regression and neural networks).

Furthermore, from a mathematical standpoint, we need to be able to analyze functions, e. g.:

– graph a function;
– find and graph the derivative of a function;
– compute the definite integral of a function.

This text addresses all of these issues to some degree. The book is intended for math and science students who have had a semester of calculus. We will approach topics from an introductory level. Because of this, we will have to exclude much of the rich theory that is available in the study of numerical methods. Our goal is to introduce students to the types of methods that are available and the basic ideas that motivate these methods. In general, there are more advanced (and more efficient) methods available than the ones we cover. However, we seek to teach the student "how" to approach a problem within the context of computing. If a student wishes to pursue a topic more deeply, we reference avenues for such further study.

In order to present the techniques and methodologies, we use the Python programming language. Thus, in addition to learning the numerical methods, students will also

https://doi.org/10.1515/9783110776645-001

learn how to program using Python. It is a powerful language that is available to everyone at no cost (since it is open-sourced). The text begins by discussing some of the fundamental tasks that we must be able to accomplish using the programming language. Such tasks include:

- arithmetic with Python;
- defining and graphing a function;
- manipulating matrices.

Once these fairly basic ideas are discussed in the context of the Python language (Chapters 2, 3, and 4), we then move on to discuss more advanced numerical methods and apply them in scientific settings.

## Data files

The data files that are used within the text may be obtained from the following address:

https://www.degruyter.com/document/isbn/9783110776645/html

or by contacting the author at wmiles@stetson.edu.

# 2 The basic operations in Python

## 2.1 Obtaining Python

The Python programming language may be obtained via several sources. A quick web search will indicate many websites from which students may download and install the language. Since there are so many possibilities, we do not attempt to define the installation process. We rely on the student to find and install the language. Students should be sure to install a version of Python 3 because Python 2 will not be supported in the future. For this text, we used Python version 3.8, and the reader should be aware that the language is continually being updated, and sometimes methods and syntax are modified or removed as new versions of the language are developed. Sites such as Python.org and ActiveState.com have been reliable sources as of the writing of this text (November 2022). In addition to the language, most students will find it helpful to have a language editor. We have used PyCharm, from www.jetbrains.com, as the editor while writing the code for this book, but there are many nice editors available. Students should install the Python language and the language editor of choice (usually, Python is installed first, then the editor) before proceeding with the following material.

## 2.2 Addition, subtraction, multiplication, and division

Once Python has been installed, our first goal is to be able to perform simple arithmetic operations. This is relatively straightforward. So, let's open a new file. We will use PyCharm as the editor, but you may choose to use a different editor. We recommend that the editor recognize Python as the language. Features such as command highlighting and completion and automatic indentation are very helpful.

Most of the arithmetic operations are as one might expect. For example, to add 3 and 5, we simply type '3 + 5'. Examples of other operations are similar:

| | |
|---|---|
| Addition: | 3 + 5 |
| Subtraction: | 3 − 5 |
| Multiplication: | 3 * 5 |
| Division: | 5/3 |

**Code:**

```
1   # Chapter 2:
2   # 2.1,Arithmetic
3
```

https://doi.org/10.1515/9783110776645-002

```
4   # Addition, Subtraction, Multiplication, Division
5
6   3+5
7   3-5
8   3*5
9   5/3
```

If we were to run a program with these commands, it would appear that nothing has happened. We get the message "Process finished with exit code 0." Python actually did the operations requested, but we failed to ask to see the results. The simplest way to see a result is to use the *print* command. To create the program to show these results, we proceed as follows.

We need to write the commands in a *program* or *script*. A *program* is the list of commands to be executed sequentially by the language (in this case, Python) compiler or interpreter. There is a difference between a complied language and an interpreted language, but, for our purposes, that difference is not of importance. We write the commands in the language editor.

When we open PyCharm for the first time, we are asked to either open an existing project or create a new one. We will create a new project called *NumMethodProject*. The student can choose a different name if desired. This generates an editor window that looks like the following.

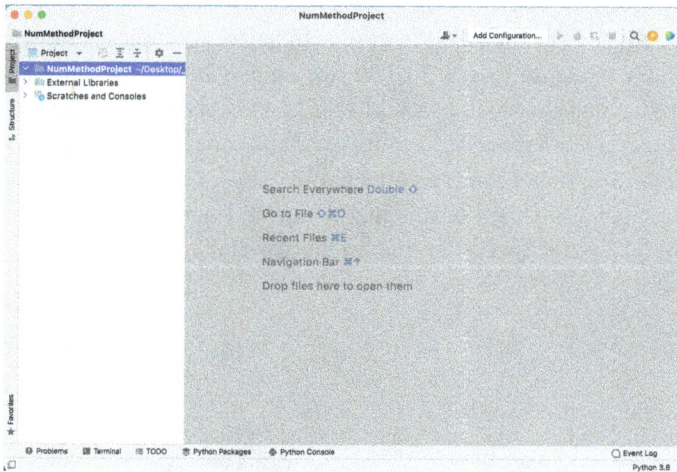

We then create a new Python file using *File → New → Python File.*

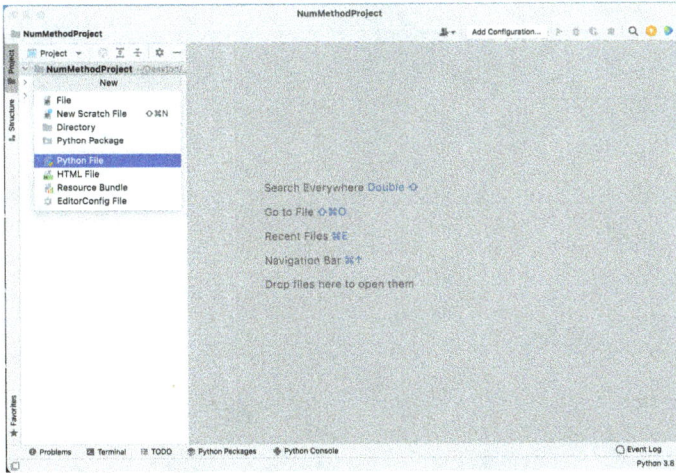

We named the file *NumMethod.*

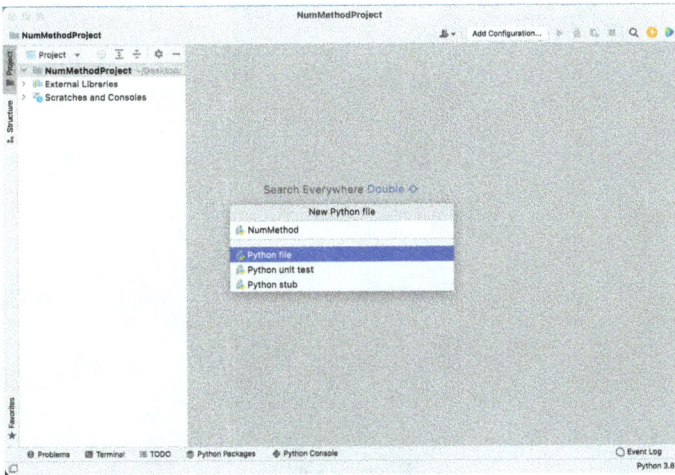

Finally, we can type the commands in the program.

Now, we can run the code using the ▷.

Then, we choose the program we wish to run.

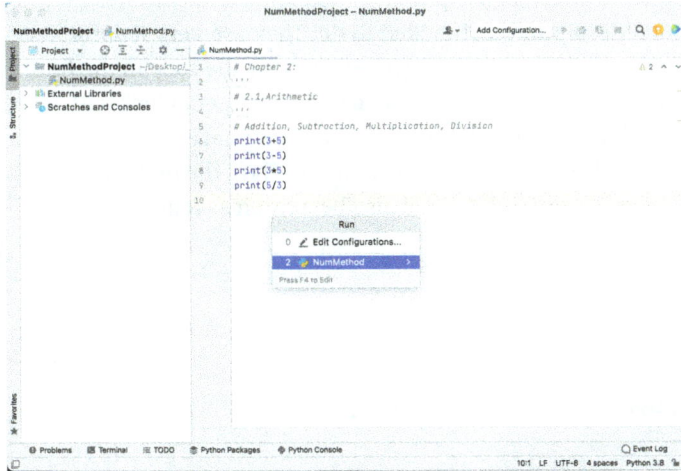

As a result, the program runs and the output is displayed in lower frame of the editor window. The exit code of '0' indicates that the program was executed (ran) with no errors. If a code is given, then the code corresponds to a specific error condition that indicates we need to correct our code in some way.

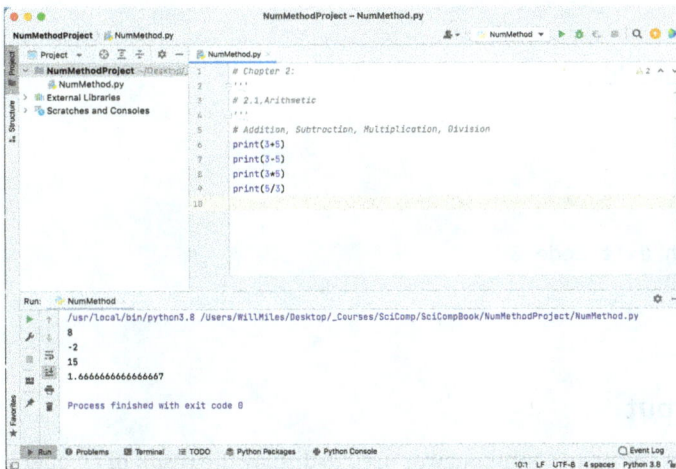

Screenshots are shown for this first example. However, we will list the code and the output separately without actually showing screen shots henceforth. We can create as many files as desired (using the same method as just shown) within the project.

Note that we can provide comments in Python by beginning the line with a # character. A *comment* is a line that is not executed when the program is run, but, instead, supplies information and context to the programmer. Usually, comment lines indicate what a program (or part of a program) is intended to do. They are very helpful in explaining the code to other programmers who may use or modify the code in the future. If we wish to have several lines as comments, we can begin a comment block with three single quotation marks. All lines that occur after the three quotes are considered to be comments until another set of three quotes are encountered. Thus, the first four lines of the preceding program (the comment lines) could also be written as follows:

```
1    '''
2    Chapter 2:
3    2.1,Arithmetic
4
5    Addition, Subtraction, Multiplication, Division
6    '''
```

## 2.3 Powers

Powers are indicated slightly differently than we might expect. To raise a number to a power, we use ** as the operator. Thus, $5^3$ is expressed as 5**3.

**Code:**
```
1    print(5**3)
```

The output should look like this.

**Output:**
```
125

Process finished with exit code 0
```

**See Exercise 1.**

## 2.4 Displaying output

In the previous examples, the print command was used to display the result of an operation. However, there are many times when we would like to display output in a particular way. For example, consider the following script:

```
1    print(1/3)
```

**Output:**
```
0.3333333333333333

Process finished with exit code 0
```

While the output is correct, one rarely needs one third expressed to 13 decimal places. Thus, we wish to format the output to be more visually pleasing (without affecting the value of the result). In addition, it is common to have multiple results that need to be displayed.

We can modify the print statement to print more than one piece of information by using commas between the separate items to be printed. Note, in the following example, the first object to be printed is a literal string (i. e., a list of characters), '3+5 ='. The string is printed as it appears, followed by the value of 3 + 5.

**Code:**
```
1   print('3+5 =',3+5)
```

**Output:**
```
3+5 = 8

Process finished with exit code 0
```

We can format output so that it is more readable by using the *.format()* function within Python. There are many options that may be applied to the *.format* method, but the most common option for us is one that allows us to fix the number of decimal places to be displayed. Suppose we wish to print 2/3 to four decimal places. We could do so with the following code.

**Code:**
```
1   print('The value of 2/3 to four decimal places is {:.4f}. '.format(2/3))
```

The output is thus:

**Output:**
```
The value of 2/3 to four decimal places is 0.6667.
```

In this line of code, the string that we wish to display is enclosed in single quotes. The braces, along with the colon, indicate that an argument will be supplied in the format section, and the .4f forces the floating-point number that is generated to be displayed out to four decimal places. We can have more than one argument. Suppose that the radius of a circle is four. The following example prints the radius and the area of the circle, using various format options. Recall that the area of a circle is given by $A = \pi r^2$. We use 3.14 to approximate $\pi$.

**Code:**

```
1  print('The radius is {}, and the area is {:.3f}.'.format(4.0, 3.14*4**2))
```

**Output:**

```
The radius is 4.0, and the area is 50.240.
```

In this format, the value of the radius is the first value in the format list. Thus, it is as-sociated with the first pair of braces in the string portion of the statement. The radius is printed as given (4.0) in the format list because no format specification is included within the braces. The area is the second value in the format list. The area is displayed using three decimal places. We contrast this format statement with the following one to further demonstrate the use of the *.format* structure.

**Code:**

```
print('The radius is {:.2f}, and the area is {:.5f}.'.format(4.0, 3.14*4**2))
```

**Output:**

```
The radius is 4.00, and the area is 50.24000.
```

In this case, two decimal places are displayed for the radius because the format specifica-tions now include *.2f* while five decimal places are used for the area, as also stipulated.

**See Exercise 2.**

## 2.5 Exercises

1.  Use Python to evaluate the following expressions:
    (a) $\frac{4.1^2 - 4^2}{.1}$
    (b) $(3 + 2)^3 (5 - 1)^4$
2.  Recall that the volume of a sphere is given by $V = \frac{4}{3}\pi r^3$. For a sphere of radius 4.23 cm, use the format structure to output the sentence: "The radius is 4.23 and the volume is xx.xxx." where the volume is computed according to the formula (and replaces the xx.xxx). Let $\pi$ be approximated by 3.14.

# 3 Functions

## 3.1 Exponentials, logs, and trig functions

We have seen how to raise numbers to powers with the ** operator. What about calculating something like $e^3$? There are multiple ways to do this, but each of them requires that we ask Python to use a set of tools that is not included in the core Python language. Such a set of tools is called a *library* or *package*. To access the package, we use the *import* command. There are at least two packages that contain the tools we need to compute $e^3$: the *math* package and the *numpy* package. Within the *math* package, there is a method or object called *e* that represents the number *e* as we know it. To access the object, we use *math.e*. The *import* statement is used to make the package available to Python. See the following code.

**Code:**
```
1   import math
2   print(math.e)
3   print(math.e**3)
```

This will display the following:

**Output:**
```
2.718281828459045
20.085536923187664

Process finished with exit code 0
```

So, now we have access to the value of *e*, and we can use it as we see fit. We can obtain the value of $e^3$ with *math.e**3*. The *math* package also contains a method called *.exp* which stands for exponential. Thus, *math.exp(3)* would also give us $e^3$ as shown here.

**Code:**
```
1   import math
2   print(math.e)
3   print(math.e**3)
4   print(math.exp(3))
```

**Output:**
```
2.718281828459045
20.085536923187664
20.085536923187668

Process finished with exit code 0
```

https://doi.org/10.1515/9783110776645-003

Alternatively, we could use a library called *numpy* instead of the math library. Numpy has an method called *exp* that will compute $e^x$ with the command *numpy.exp(x)*. Thus, $e^3$ would be found with the following code.

**Code:**

```
1  import numpy
2  print(numpy.exp(3))
```

**Output:**
```
20.085536923187668

Process finished with exit code 0
```

The log functions are similar. For example, to compute ln(5), we could use either *math.log(5)* or *numpy.log(5)*, depending on which library we choose to import. Note that, in both of these packages, "log" indicates "ln" instead of the base 10 log. In our experience, we tend to use numpy more often than the math library.

Both Numpy and the math package also provide the trigonometric functions. The angle measures are assumed to be in radians. Suppose we wish to print a table of sine and cosine values for the standard angles: $0$, $\frac{\pi}{6}$, $\frac{\pi}{4}$, $\frac{\pi}{3}$, and $\frac{\pi}{2}$.

**Code:**

```
1   #print a table of trig values
2   import numpy
3   print("angle |{:>5} |{:>5} |{:>5} |{:>5} |{:>5} |".format(\
4       '0','\u03C0/6','\u03C0/4','\u03C0/3','\u03C0/2'))
5   print('------------------------------------------')
6   print("cos(x)|{:.4f}|{:.4f}|{:.4f}|{:.4f}|{:.4f}|".format(\
7       numpy.cos(0),numpy.cos(numpy.pi/6),numpy.cos(numpy.pi/4),\
8       numpy.cos(numpy.pi/3),numpy.cos(numpy.pi/2)))
9   print("sin(x)|{:.4f}|{:.4f}|{:.4f}|{:.4f}|{:.4f}|".format(\
10      numpy.sin(0),numpy.sin(numpy.pi/6),numpy.sin(numpy.pi/4),\
11      numpy.sin(numpy.pi/3),numpy.sin(numpy.pi/2)))
```

**Output:**

```
angle |   0 | π/6 | π/4 | π/3 | π/2 |
------------------------------------------
cos(x)|1.0000|0.8660|0.7071|0.5000|0.0000|
sin(x)|0.0000|0.5000|0.7071|0.8660|1.0000|

Process finished with exit code 0
```

**Figure 3.1:** Table of trigonometric values.

There are a few things to learn from this short code. First, line 3 makes use of a string formatting argument to make the width of the output fixed. The notation { :>5} indicates a string field of a width of five characters. This makes possible a vertical alignment of the table since more lines are printed. To print the $\pi$ symbol, we used the code \u03C0. The codes for the symbols may be found easily online. At the time of this writing, many useful codes were found here:

https://pythonforundergradengineers.com/unicode-characters-in-python.html

The \ at the end of lines 3, 6, 7, 9, and 10 indicate that the command continues to the next line. This allows for more readable code should the lines become lengthy. Finally, one can see the numpy.sin(), numpy.cos(), and the numpy.pi methods used throughout the code.

**See Exercise 1.**

## 3.2 Variables

When using a computer language, we need to be able to store values for future use. We use *variables* to do this. In mathematics, we are accustomed to using variables like $x$ and $y$. In programming, it is common to use variable names that are more descriptive. Variable names may not begin with a number, and, while the underscore character may be used, most special characters are not allowed in the name. Otherwise, variables can be named almost anything. For example, in mathematics we would tend to denote height as a variable by using $h$ or $h(x)$, but, in Python (or most computer languages), one would likely use the entire word *height* as the variable name. Hence, there is no confusion as to what the variable represents. In that sense, programmers use variable and function names in a more explicit and meaningful way than mathematicians. In this text, there are three main types of variables used, namely,
- floating point
- integer
- string

A floating-point variable is a variable that can hold a decimal number while an integer variable expects only a pure integer. It is important to use floating-point variables when needed.

If a variable is assigned an integer value (no decimal point), then Python assumes the variable is an integer variable. In Python 3, if an arithmetic operation involving integer variables does not naturally result in an integer, the then Python will convert the result to floating-point type. Such conversion is not true of all programming languages, so one must be mindful of the types of variables that are being used.

Consider the following:

**Code:**

```
1   r = 4
2   print(type(r))
3   q = 3
4   print(type(q))
5   a = r/q
6   print(a)
7   print(type(a))
```

**Output:**

```
<class 'int'>
<class 'int'>
1.3333333333333333
<class 'float'>
```

Note that r, q, and a are all variables in this code. We use the type command to see what kind of contents are stored in the variable. From this, we see that, while r and q are both integer-type variables, the result of r/q results in a non-integer. Fortunately, Python recognizes this and assigns a to be a floating-point variable.

A variable that holds characters (or text) is called a *string* variable. Some example code follows.

**Code:**

```
1   x = 'Will Miles'
2   print(x)
```

In this code, x holds the string Will Miles. The string must be enclosed by either single or double quotes. When we run the code, the following output is given.

**Output:**

```
Will Miles
```

There are many methods that are available for use with strings. Perhaps surprisingly, we can add strings and multiply them by counting numbers. The results of these operations are shown via the following examples.

**Code:**

```
1   fname = 'Will'
2   lname = 'Miles'
3   name = fname+lname
4   print(name)
```

**Output:**
```
WillMiles
```

Addition of strings results in the second string being attached to the end of the first string. Such a combination is called a concatenation. Notice that no space is added between the strings that are being added. If we want a space, we could either add a space to fname, or we could add a space explicitly in the expression:

```
name = fname + ' ' + name
```

Multiplying a string by a nonnegative integer creates the specified number of copies of the string concatenated as shown next.

**Code:**
```
1   fname = 'Will'
2   name = 4*fname
3   print(name)
```

**Output:**
```
WillWillWillWill
```

Strings are officially *lists* of characters. Lists will be discussed in more detail later in the text, but we can access portions of a string variable as follows. Consider the code that follows.

**Code:**
```
1   coursename = 'Scientific Computing'
2   print(coursename[3])
3   print(coursename[0:10])
4   print(coursename[11:])
```

**Output:**
```
e
Scientific
Computing
```

Notice that coursename[3] gives the fourth character of the string. This is because Python begins counting at 0. When we write coursename[0:10], we are asking for a range within the list, starting at the $0^{th}$ element and ending with the $9^{th}$ element. So, [0:10] indicates from 0 to 10, not including 10. Similarly, coursename[11:] requests the elements of coursename beginning with element 11 and continuing until the end of the list (or string in this case). This colon notation allows us to access pieces of the string or list as we wish.

Another common task associated with strings is the need to find a particular string inside of another string. This is accomplished with the .find attribute. We use the following structure:

*variable name*.find('*characters to search for*')

The following code checks to see if the string 'Comp' is inside the string defined by coursename. It then attempts to find the string 'not' inside of coursename.

**Code:**
```
1   coursename = 'Scientific Computing'
2   a = coursename.find('Comp')
3   print(a)
4   b = coursename.find('not')
5   print(b)
```

**Output:**
```
11
-1
```

When a string is found, the index of the beginning of the substring is returned. That is, 'Comp' begins at the eleventh[th] element of coursename. However, if the string is not found, the method returns a value of −1. When control structures are introduced later, one will see how this information can be used. The .find method is case sensitive. That is, if we were to search for 'comp' instead of 'Comp', the method would return a −1. If one wishes to remove the case sensitivity, we could use the string method .upper to convert the string to the upper case before searching the string.

**Code:**
```
1   #use upper case to ignore case sensitivity
2   coursename = 'Scientific Computing'
3   # convert string to upper case
4   Ucourse = coursename.upper()
5   print(Ucourse, coursename)
6   #search for the uppercase of 'comp'
7   a = Ucourse.find('comp'.upper())
8   print(a)
```

**Output:**
```
SCIENTIFIC COMPUTING Scientific Computing
11

Process finished with exit code 0
```

Examining the previous code, we see that the contents of coursename are converted to all uppercase letters in line 4. When the converted string, Ucourse, and the original name are printed in line 5, we see that the original variable contents are unchanged. In line 7, the uppercase string is searched for the uppercase substring. Thus, since all letters are uppercase, there is no case sensitivity.

There are many other string methods available. Essentially, if you want to do something with strings, your first step should be to Google what you want with 'Python' included in the search terms. It is likely that a method is already included among the returned content.

**See Exercises 2–4.**

## 3.3 Defining and using mathematical functions

We have seen that Python has many mathematical functions (including trigonometric, exponential, and logarithmic functions) available via the math and numpy libraries. However, it is often the case that we would like to define our own functions and be able to access them in a convenient fashion. We can do this in Python by using the def structure. Suppose we wish to define and use the function $f(x) = 3x^2 - 2x + 1$. We would begin by defining the function with the following.

**Code:**
```
1   def f(x):
2       y = 3.0*x**2-2.0*x+1.0
3       return y
```

The code starts with the def command followed by the name of the function. We can choose any name for the function. Generally, like variable names, function names tend to indicate the purpose of the function. In this case the function name is f. Following the function name is a list of inputs that the function will need in order to compute its value or perform its task. In this example, the function f requires a value for x in order to calculate the value of the function. In this example the value of the function is stored in the variable y. Finally, we must return the value of y. The indentation after the *def* declaration is important. All lines of code that are indented under the def are part of the definition of the function. Once the indentation ends, the function is complete. The return is required because defined functions have what is called *local scope*. This means that variables and values used within the defined function are known only to that function. Hence, if the main program refers to variables that are defined within the function, an error will likely occur. For example, consider the following.

**Code:**

```
1   #defining functions
2   x = 3
3   def f(x):
4       y = 3.0*x**2-2.0*x+1.0
5       return y
6
7   print(y)
```

The program begins by defining x to be 3. The function f is then defined as before. However, when the code attempts to print y, an error is encountered.

**Output:**

```
Traceback (most recent call last):
  File "/Users/WillMiles/Desktop/_Courses/SciComp/SciCompBook/BookCode/
        BookCodeChap3.py", line 75, in <module>
    print(y)
NameError: name 'y' is not defined

Process finished with exit code 1
```

This is because the y variable that is defined in the function definition is restricted to just that function. Hence the "mainline" program does recognize y as being a defined variable. This local scope allows us to reuse variable names if we wish, i. e., we could still define a variable y in the mainline without affecting the variable used inside the function definition. Furthermore, defining x to be 3 has no meaning to the function. The variable x is defined outside of the function definition; thus, the function does not know the value of x.

To use (or call) the function, we use the function name and supply the necessary arguments. For example, one can evaluate the function as usual, using $f(x)$ notation. For example, if we wished to know the value of $f$ when $x = 3$, we would calculate $f(3)$. We can do this in Python as shown in the following.

**Code:**

```
1   #defining functions
2   def f(x):
3       y = 3.0*x**2-2.0*x+1.0
4       return y
5
6   y = f(3)
7   print('f(3)=',y)
```

The output displays as follows.

**Output:**
```
f(3)= 22.0
```
```
Process finished with exit code 0
```

As another example to demonstrate the scope issue regarding functions, consider the following.

**Code:**
```
1   #defining functions
2   def f(x):
3       y = 3.0*x**2-2.0*x+1.0
4       return y
5
6   y = 10
7   print('f(3)=',f(3))
8   print('y=',y)
```

**Output:**
```
f(3)= 22.0
y= 10
```
```
Process finished with exit code 0
```

Notice that y is used in both the function and the mainline program. When the function is called, it does not change the value of y that exists in the main.

**See Exercises 5–6.**

## 3.4 Getting input from the keyboard

Often, we wish to have a program ask for input from the user. For example, we may wish to enter the value of the radius of a circle and have a Python compute the area of the circle. We can accomplish this input via the input command. The command uses the following syntax:

```
variable_name = input('prompt string')
```

Input is received as a string. Thus, if we wish to use the input as if it were numeric, we must convert the string to a useable number. We can do this by simply taking the input and applying the desired type to it. Consider the following code that prompts for the radius of a circle to be entered, computes the corresponding area, and displays the results.

**Code:**

```
1  import numpy as np
2  radius_str = input('Enter the radius: ')
3  #convert the radius to a floating point value
4  radius = float(radius_str)
5  #compute the area, A = pi*r^2
6  area = np.pi*radius**2
7  print('The area of a circle with radius {:.3f} cm is {:.3f}\
8    square cm'.format(radius, area))
```

The prompt is issued in line 2 of this code, and the number that is entered (as a string by default) is converted to a floating-point variable in line 4. There are some issues that we should recognize in this small code. First, if the user enters something other than a number, an error is likely to occur in line 4 when it tries to convert the entered text into a number. There are ways to check to see if the input is valid, but we will not address that at this time. For our purpose, we wish to develop code that will help us to solve problems. Hence, we simply recognize that we need to be careful when entering data. Also, in line 1, notice that numpy was imported as np. This allows us to type np instead of numpy when we access elements of the numpy package, as we did in line 6. A sample output is given in the following.

**Output:**

```
Enter the radius: 4.2
The area of a circle with radius 4.200 cm is 55.418 square cm

Process finished with exit code 0
```

**See Exercises 7–8.**

## 3.5 Graphing functions

In order to work with functions we frequently wish to see the graph. Thus, we turn our attention to the task of producing graphs of functions using Python. To do this, we make a list of *x* values and use the defined function to compute the corresponding list of *y* values. We use another library called *matplotlib* to generate the graph. The plotting methods are in a sub-module of matplotlib called pyplot. Hence, we can import just that part of the library to save some memory and to reduce the amount of typing needed to call up the methods.

To make a list of *x* values, Python offers several alternatives. The standard *list* (or array) is simply a list of items enclosed in square brackets and separated by commas. In the following code, x and y are both standard lists in Python. Note that x is a list of integers, while y is a list of mixed type: some integers, a floating-point number, and a three strings.

**Code:**
```
1  #Python lists
2  x = [1,2,3,4]
3  print('x=',x)
4  y = [1,'a',3.14,'c','will']
5  print('y=',y)
```

When the code is run, the following is displayed.

**Output:**
```
x= [1, 2, 3, 4]
y= [1, 'a', 3.14, 'c', 'will']

Process finished with exit code 0
```

Just as we did with strings, we can access lists using square brackets. So, x[1] will be 2 (again, because Python begins counting at 0), and y[4] = 'will'. We can also use the colon notation to access parts of the lists as we previously did with strings.

So, suppose we wish to plot the function $f(x) = x^2$ between $x = 0$ and $x = 5$. To do this in Python, we need a list of *x* values and a corresponding list of *y* values. So, let's define the list for *x* to be

```
x = [0,1,2,3,4,5].
```

Then the corresponding list for *y* would be

```
y = [0,1,4,9,16,25].
```

Now we want to plot these points and connect them. This is where we will use the *matplotlib* package. The code to plot the points follows.

**Code:**
```
1  #graphing functions
2  import matplotlib.pyplot as plt
3  x = [0,1,2,3,4,5]
4  y = [0,1,4,9,16,25]
```

```
5   plt.plot(x,y)
6   plt.show()
```

In line 1, we import the plotting functions that are included in the matplotlib package. Notice that we imported the desired module and named it plt. This allows us to refer to the library using this abbreviated name, saving us some typing. Lines 2 and 3 define the points that are on the graph of the function to be plotted. Line 4 actually creates the plot but does not display the plot. Finally, line 5 shows the plot. In lines 4 and 5, the 'plt' references the library of plotting functions that was imported. We use the dot notation to indicate which method from that library we wish to use. When this is run, we should see the following plot.

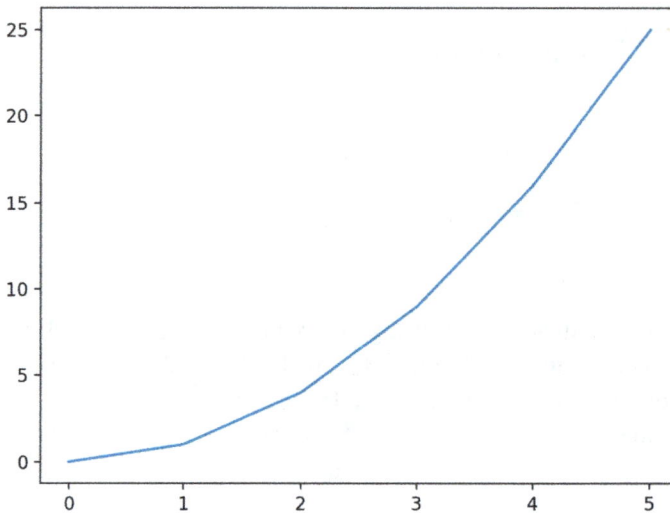

It is important to remember to include the plt.show(). Without it, it will appear that nothing has been calculated. We can also include many other attributes of a graph. We can place titles on the axes or on the entire graph, we can change the plotting style, and we can add a grid if desired. Consider the code below.

**Code:**
```
1   #graphing functions
2   import matplotlib.pyplot as plt
3   x = [0,1,2,3,4,5]
4   y = [0,1,4,9,16,25]
5   plt.plot(x,y,'b*-',label='f(x)=x^2')
6   plt.xlabel('x-axis')
7   plt.ylabel('y-axis')
```

```
8    plt.title('Big Title')
9    plt.grid()
10   plt.show()
```

The code should generate the plot that follows.

**Output:**

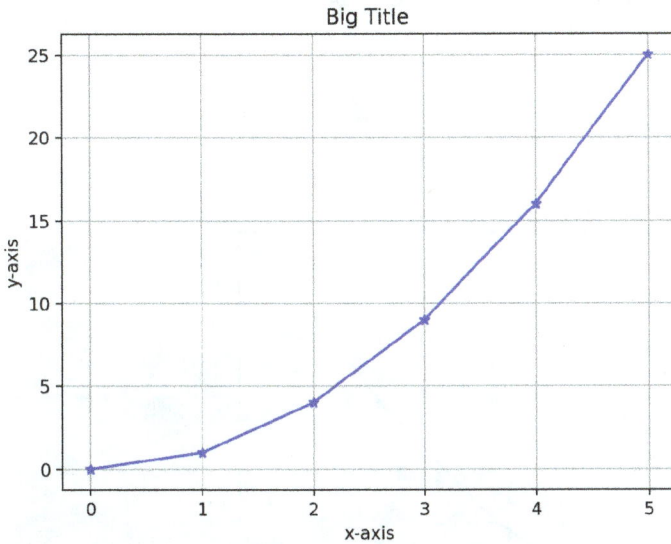

There are many new attributes included in the previous code. In line 5, some plotting options are included in the plot command. The string `'b*-'` is a called a format string and indicates that a blue line is to be used to connect data points. Further, the actual data points are to be marked with an asterisk. The `label` attribute in line 5 allows one to apply a description, $f(x) = x\^2$, to this plot so that we can include a legend if desired. Lines 6 and 7 indicate how to add axis titles to the plot, and line 8 allows for an overall title. Finally, line 9 displays a graphing grid for the plot.

We can also include another plot on the same set of axes and add a legend to the graph. Suppose we wish to add a graph of the derivative $f'(x) = 2x$ to the current plot. We can use the same list of $x$ values, but we will need a new list of $y$ values. In the following code, we use $z$ to store the list of $y$ values that corresponds to $2x$. Thus, $z=[0,2,4,6,8,10]$. Then, another plot command is executed.

**Code:**

```
1    #graphing functions
2    import matplotlib.pyplot as plt
3    x = [0,1,2,3,4,5]
```

```
4    y = [0,1,4,9,16,25]
5    z = [0,2,4,6,8,10]
6    plt.plot(x,y,'b*-',label='f(x)=x^2')
7    plt.xlabel('x-axis')
8    plt.ylabel('y-axis')
9    plt.title('Big Title')
10   plt.grid()
11   plt.plot(x,z,'b--',label="f'(x)=2x",c='0.45')
12   plt.legend()
13   plt.show()
```

And the graph now looks like the following.

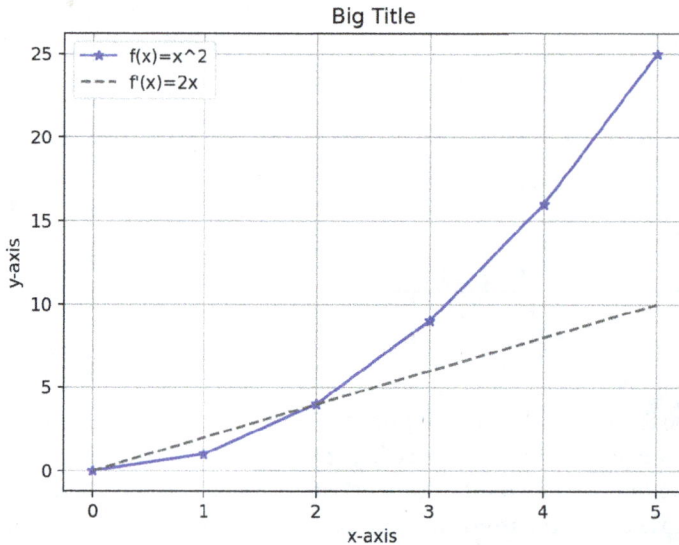

In line 11, we used another kind of color attribute. The attribute c='0.45' allows us to plot in a gray scale. The variable c can take on any number between 0 and 1 where 0 implies the color is black and 1 is white. Values in between set the color to a level of gray accordingly. Thus, the curve is to be plotted in a shade of gray using a dashed line with no data markers. The legend is placed in line 12. A full list of plotting options can be found in the documentation for matplotlib:

https://matplotlib.org/stable/api/_as_gen/matplotlib.pyplot.plot.html

**See Exercises 9–10.**

While the previous example is successful in plotting $y = x^2$ between $x = 0$ and $x = 5$, the graph has noticeable corners, and we had to manually type in all of the points instead of using the actual function to calculate the $y$ values. Thus, we'd like to find a way to include a great many more points on the graph (so that the graph is smoother), and we want to define the function and use it within Python instead of calculating values by hand. Suppose we would like to use $x$ values that are spaced 0.1 units apart. So, we would have a list that looks like

$$x = [0, 0.1, 0.2, 0.3, \ldots, 5].$$

We could type the list of 51 numbers, but that kind of defeats the purpose of using a computer. Fortunately, the *numpy* library includes a method that will do exactly this. We can generate a *numpy list* using a method called *arange(start, stop, step)*. The method takes three arguments:
- **start**: the first number in the desired range of numbers;
- **stop**: the number at which the list will end (it is not included in the list);
- **step**: the interval between each pair of numbers in the list

Consider the following code.

**Code:**
```
1  import numpy as np
2  x = np.arange(0,5,1)
3  print('x=',x)
```

In line 1, we import the numpy library as *np*. We then assign a list of values to x, starting with 0, ending at 5, separated by 1. However, note that 5 is not included. The program produces what looks like this.

**Output:**
```
x= [0 1 2 3 4]

Process finished with exit code 0
```

If we wanted 5 to be included in the list, then the "end" value would have to be larger than 5. Let's change 5 to 5.1. Now x is as shown below.

**Output:**
```
x= [0. 1. 2. 3. 4. 5.]

Process finished with exit code 0
```

Now, 5 is included in the list. Also, decimal points have now been used in all the numbers. That's because, when the Python interpreter encountered "5.1," it assumed that floating-point numbers were now allowed and expected.

So, to get our list of 51 points between 0 and 5, we would need $x$ = np.arange(0,5.1,0.1). In general, if we wish to construct a list that starts with a and ends with b (including b) with a step of dx, then we should add the step size to the end value. So, the command would look like this:

```
x = np.arange(a,b+dx,dx)
```

Now we wish to fill another list with the associated $y$ values. To do so, we can define the function of interest and use it to evaluate the function at all of the values in the $x$ list. The code to do this is given next.

**Code:**

```
1   import numpy as np
2   def f(x):
3       y = x**2
4       return y
5
6   x = np.arange(0,5.1,0.1)
7   print('x=',x)
8   y = f(x)
9   print('y=',y)
```

While we can define functions almost anywhere in the code, it is common to put function definitions at the top of the code. That way, we can be sure that all the functions are defined before the logic of the program begins. Lines 2–4 define the function $f(x) = x^2$. Line 6 sets up the list of $x$ values, and line 7 prints the list. We will want to eliminate the print statement once we know things are working because the list is long and uses many lines to display. In line 8, the list of associated $y$ values is constructed. Note that we use the defined function $f(x)$ to evaluate $x^2$ for all of the $x$ values at once. Line 9 prints the list of $y$ values just to make sure the list is filled correctly. Again, we will eliminate the print statement in the future. When the code is run, the following is generated.

**Output:**

```
x= [0.   0.1 0.2 0.3 0.4 0.5 0.6 0.7 0.8 0.9 1.   1.1 1.2 1.3 1.4 1.5 1.6
 1.7 1.8 1.9 2.   2.1 2.2 2.3 2.4 2.5 2.6 2.7 2.8 2.9 3.   3.1 3.2 3.3 3.4
 3.5 3.6 3.7 3.8 3.9 4.   4.1 4.2 4.3 4.4 4.5 4.6 4.7 4.8 4.9 5. ]
y= [0.000e+00 1.000e-02 4.000e-02 9.000e-02 1.600e-01 2.500e-01 3.600e-01
 4.900e-01 6.400e-01 8.100e-01 1.000e+00 1.210e+00 1.440e+00 1.690e+00
 1.960e+00 2.250e+00 2.560e+00 2.890e+00 3.240e+00 3.610e+00 4.000e+00
 4.410e+00 4.840e+00 5.290e+00 5.760e+00 6.250e+00 6.760e+00 7.290e+00
 7.840e+00 8.410e+00 9.000e+00 9.610e+00 1.024e+01 1.089e+01 1.156e+01
```

```
1.225e+01 1.296e+01 1.369e+01 1.444e+01 1.521e+01 1.600e+01 1.681e+01
1.764e+01 1.849e+01 1.936e+01 2.025e+01 2.116e+01 2.209e+01 2.304e+01
2.401e+01 2.500e+01]
```

Process finished with exit code 0

By inspection, we see that the values for the *y* list are, in fact, the squares of the values in the *x* list. Hence, the function is working correctly. So, we now have 51 paired *x* and *y* values. In this fashion, we could build lists with as many points as desired. The reader is encouraged to modify the code to build a list of 100 points.

Note that numpy may use scientific notation to print numbers that are generated by floating-point arithmetic. Personally, I do not like this, so I frequently change the default print option with the following command:

```
np.set_printoptions(precision=3,suppress=1,floatmode='fixed')
```

The *precision=3* argument indicates that three decimal places are displayed. The *suppress=1* argument suppresses the use of scientific notation, and the *floatmode='fixed'* option causes all numbers to display all decimal places, even if there are redundant zeros. You do not need this command. I use it because I do not like the look of scientific notation, especially if the numbers are not that big.

Okay, now we can plot the function the same way we did earlier, using the matplotlib functions with the *x* and *y* lists.

**Code:**

```
1   import numpy as np
2   import matplotlib.pyplot as plt
3   np.set_printoptions(precision=3,suppress=1,floatmode='fixed')
4   def f(x):
5       y = x**2
6       return y
7
8   x = np.arange(0,5.1,0.1)
9   y = f(x)
10  plt.plot(x,y)
11  plt.xlabel('x-axis')
12  plt.ylabel('y-axis')
13  plt.title('f(x) = x^2')
14  plt.grid()
15  plt.show()
```

We can see that the resulting graph is much smoother than the previous graph.

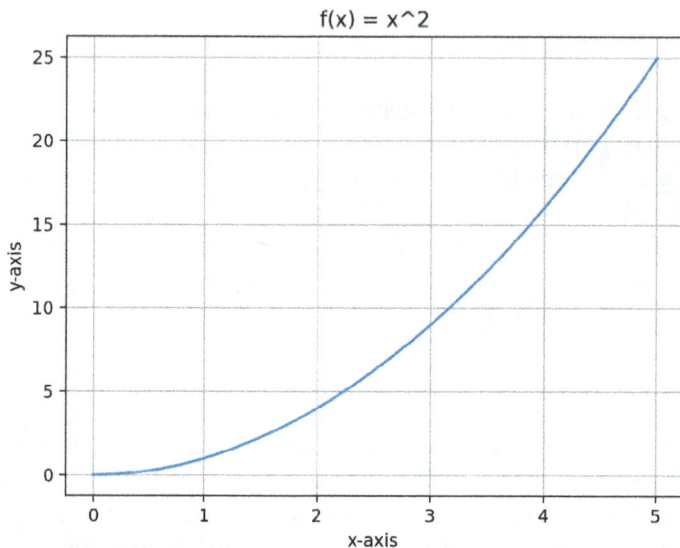

f(x) = x^2

We should also note that the figure window that is generated by matplotlib is an interactive window. We can save the graph to a file or edit portions of the graph as indicated by the tool bar at the bottom of the plot. Finally, we may want to plot multiple graphs in separate windows. We use the .figure() method to accomplish this. In the following code, we plot $f(x) = x^2$ in one window and $g(x) = \sin(x)$ in a second window. See the modified code that follows.

**Code:**

```
1   import numpy as np
2   import matplotlib.pyplot as plt
3   np.set_printoptions(precision=3,suppress=1,floatmode='fixed')
4   def f(x):
5       y = x**2
6       return y
7
8   x = np.arange(0,5.1,0.1)
9   y = f(x)
10  z = np.sin(x)
11  plt.plot(x,y)
12  plt.grid()
13  plt.figure()
14  plt.plot(x,z)
```

```
15    plt.grid()
16    plt.show()
```

When this code is executed, two figure windows are produced: one shows $y = x^2$; one shows $y = \sin(x)$.

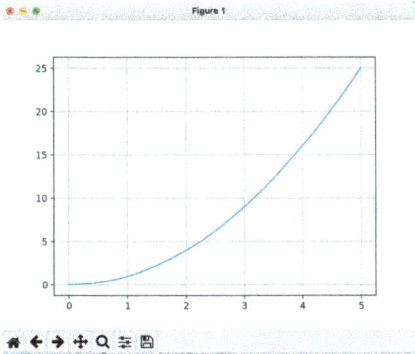

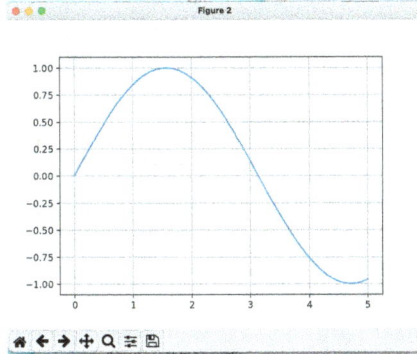

We now look at how to plot piecewise functions. Suppose we wish to graph the following function:

$$f(x) = \begin{cases} x^2 & \text{if } x \leq 0 \\ x+1 & \text{if } x > 0 \end{cases}$$

To define this function, we must be able to make decisions based on the value of $x$. In Python (and other programming languages), we do this with the *if* structure. The general form of the if structure is:

if *condition*:

       do these statements if condition is true
       executable statements

else:

       do these statements if condition is not true
       executable statements

The indentation is important because it indicates the blocks of code to be done depending on the result of the condition. For the function of interest, we would want a structure like the following.

**Code:**

```
1    #our first if statement
2    #first get an x value
3    x = float(input('Enter an x value: '))
4    #logic of the piecewise function
```

```
5   if x<=0:              #the condition is x<=0
6       y = x**2          #condition is true
7   else:                 #otherwise
8       y = x+1           #condition is false
9   print('f({}) = {}'.format(x,y))
```

Let's step through this. Lines 1 and 2 are comment lines. Python knows this because of the # at the beginning of the lines. Comments are not executed. They are there to help us explain to others what the code is doing. Line 3 allows the user to input a number. The number is converted to a floating-point value and stored in the variable x. Line 5 begins the if structure by testing to see if $x \leq 0$. Note that $\leq$ is denoted by <=. Suppose the user enters –1. Then, the condition is true. Thus, line 6 is executed, and $y = (-1)^2 = 1$. Lines 7 and 8 are not executed because the condition was true. Then, line 9 is executed to display f(-1) = 1 on the screen.

**Output:**
```
Enter and x value: -1
f(-1.0) = 1.0


Process finished with exit code 0
```

If we were to enter 2 for x, then, line 6 would not be executed because the condition would be false. So, line 8 would be executed to give $y = x + 1 = 2 + 1 = 3$. Try it.

**Output:**
```
Enter and x value: 2
f(2.0) = 3.0


Process finished with exit code 0
```

When using conditions like those commonly found in *if* structures, we often use logical operators to compare values. Thus, we need to know how to express the usual operators in the Python language. The following table shows how to write each type of operator in the appropriate way.

| Logical Operator | Python Expression |
| --- | --- |
| = | == |
| ≠ | != |
| > | > |
| ≥ | >= |
| < | < |
| ≤ | <= |

Note that the double equals (==) is used when making a comparison, while the single equals (=) is used to assign a value to a variable. We can include the if structure inside of a function definition as shown in the code below.

**Code:**
```
1   # define a piecewise function using if statements
2   # in this example, we have named the function pw
3   def pw(x):
4       #logic of the piecewise function
5       if x<=0:    #the condition is x<=0
6           y = x**2    #condition is true
7       else:       #otherwise
8           y = x+1 #condition is false
9       return y
10
11  y1 = pw(-1)
12  y2 = pw(2)
13  print('f({}) = {}'.format(-1,y1))
14  print('f({}) = {}'.format(2,y2))
```

Note that we must include another level of indentation to meet the requirements of the *def* structure. We called the new function *pw*. Also, we must include a return statement so that the value of the function can be used after it has been computed. Finally, the function is used in lines 11 and 12 and the results are displayed by lines 13 and 14. When executed, the program produces the following output.

**Output:**
```
f(-1) = 1
f(2) = 3

Process finished with exit code 0
```

Okay, so now let's try to plot the function between $x = -2$ and $x = 2$. We will use the same steps that we did in previous graphing programs:
- import matplotlib.pyplot;
- define the function;
- create numpy list of x values;
- use the function to create corresponding list of y values;
- use .plot to plot the graph of the function.

So, it seems like the following should work.

**Code:**

```
1   import numpy as np
2   import matplotlib.pyplot as plt
3   np.set_printoptions(precision=3,suppress=1,floatmode='fixed')
4   # define a piecewise function using if statements
5   # in this example, we have named the function pw
6   def pw(x):
7       #logic of the piecewise function
8       if x<=0:    #the condition is x<=0
9           y = x**2     #condition is true
10      else:        #otherwise
11          y = x+1 #condition is false
12      return y
13
14  a = -2
15  b = 2
16  n = 100
17  dx = (b-a)/n
18  x = np.arange(a,b+dx,dx)
19  y = pw(x)
20  plt.plot(x,y)
21  plt.show()
```

However, when we run this, we get the following error.

**Output:**

```
Traceback (most recent call last):
  File "/Users/WillMiles/Desktop/_Courses/SciComp/SciCompBook/BookCode/
      BookCodeChap3.py", line 163, in <module>
    y = pw(x)
  File "/Users/WillMiles/Desktop/_Courses/SciComp/SciCompBook/BookCode/
      BookCodeChap3.py", line 152, in pw
    if x<=0:    #the condition is x<=0
ValueError: The truth value of an array with more than one element is
    ambiguous. Use a.any() or a.all()

Process finished with exit code 1
```

The problem occurs when we ask the function *pw* to compute a value for each of the values in the x list. Once the if structure was introduced to the function, the function no longer knows how to evaluate a distinct value for each value in the x list. The *pw* function is really expecting just a single number (a scalar) as input. So when it encountered a list of numbers, the if structure was unable to perform the comparisons in an element-by-element fashion. To address this issue, we will "vectorize" the *pw* function. We create a new function that will do the element-by-element comparison with the following command:

```
vpw = np.vectorize(pw)
```

The function is named *vpw* (for vectorized pw). You can name it anything, but I tend to just put a 'v' in front of the existing function's name so that I can keep track of the vectorized functions if I have more than one. The revised code follows.

**Code:**

```
1   import numpy as np
2   import matplotlib.pyplot as plt
3   np.set_printoptions(precision=3,suppress=1,floatmode='fixed')
4   # define a piecewise function using if statements
5   # in this example, we have named the function pw
6   def pw(x):
7       #logic of the piecewise function
8       if x<=0:    #the condition is x<=0
9           y = x**2    #condition is true
10      else:       #otherwise
11          y = x+1 #condition is false
12      return y
13  vpw = np.vectorize(pw)
14  a = -2
15  b = 2
16  n = 100
17  dx = (b-a)/n
18  x = np.arange(a,b+dx,dx)
19  y = vpw(x)
20  plt.plot(x,y)
21  plt.show()
```

Notice that the vectorized version of pw is formed in line 13, and line 19 has been changed so that vpw is used instead of pw. The list of *x* values is created in lines 14–18. This may look like a lot of work just to get a list of values, but structuring the list like this allows us to enter the start and stop values easily. It also allows us to change the number of intervals (number of points minus 1) we wish to use with ease. The spacing is calculated by the code. So, a little extra work up front leads to more flexibility of the code in the long run. The code now runs without error and produces the following plot.

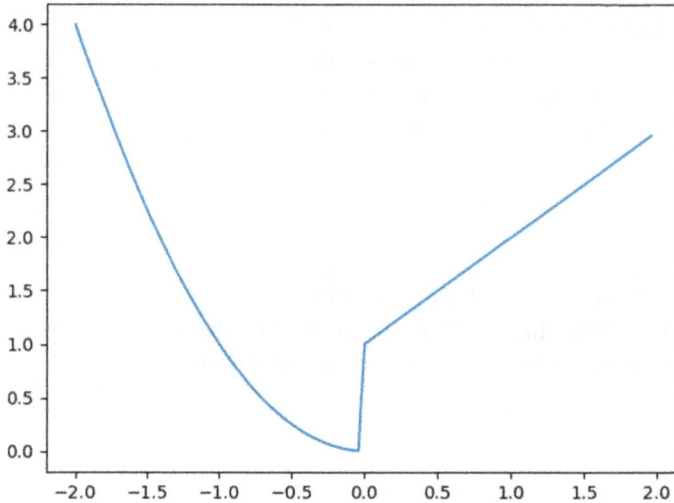

From the plot, we can see that there are still problems with the graph at the point where the jump discontinuity occurs (when $x = 0$). The graph is connected when it should display a jump from one curve to the other. One way to address this unwanted connection is to draw the graph in separate segments, i. e., draw the $x^2$ part, and then draw the $x + 1$ part, using appropriate lists of $x$ values for each part. Thus, we construct a list of $x$ values that are between –2 and 0 (including 0) and a separate list of values between 0 and 2 (excluding 0). Then, use the function to fill corresponding lists of $y$ values. Finally, plot each $x$-$y$ pair on the same graph. The code to implement this logic is below.

**Code:**

```
1    import numpy as np
2    import matplotlib.pyplot as plt
3    np.set_printoptions(precision=3,suppress=1,floatmode='fixed')
4    # define a piecewise function using if statements
5    # in this example, we have named the function pw
6    def pw(x):
7        #logic of the piecewise function
8        if x<=0:     #the condition is x<=0
9            y = x**2     #condition is true
10       else:        #otherwise
11           y = x+1 #condition is false
12       return y
13   vpw = np.vectorize(pw)
14   #set up the list for -2<x<=0
15   a = -2
16   b = 0
```

```
17    n = 50
18    dx = (b-a)/n
19    x = np.arange(a,b+dx,dx)
20    y = vpw(x)
21    plt.plot(x,y,'b')
22    #now do the second section of the function
23    a = 0
24    b = 2
25    n = 50
26    dx = (b-a)/n
27    # in this list we want to exclude the left endpoint at 0
28    # so we will use a start value that is slightly larger than 0
29    x = np.arange(a+dx,b+dx,dx)      #note, this includes 2
30    y = vpw(x)
31    plt.plot(x,y,'b')
32    plt.grid()
33    plt.show()
```

The graph that is produced looks like this.

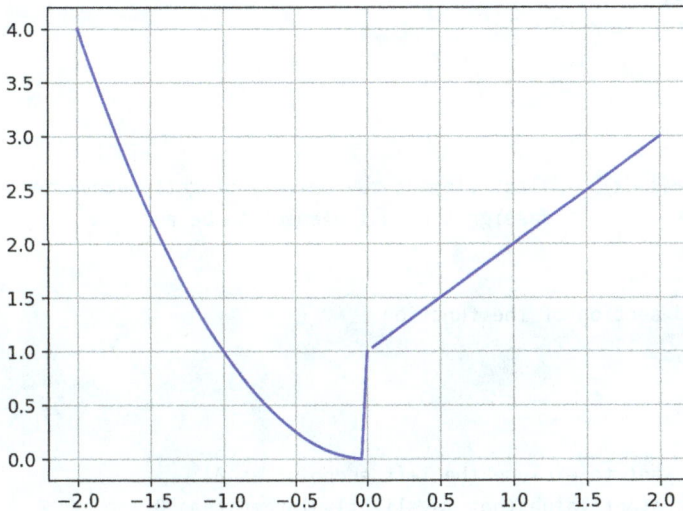

So, it seems that we did not fix the issue. We can see that the second part of the function (the linear piece) seems to be fine. It looks like there is a small gap when $x = 0$ which was accomplished in line 29 by starting the list at a+dx. So what happened in the first part of the graph? Well, because of the way a computer does arithmetic, numbers can be off by a very small amount, which is caused by computer rounding error. In this case, when

Python computed the x list, the last number in the list (which we designed to be zero) is actually verb 1.7763568394002505e-15 . While this is really close to zero, it is slightly greater than zero. Thus, the second part of the piecewise function was used to calculate the y value, which results in an obvious error in the graph. To avoid this rounding error, we can simply assign the last element of the list to be what we want it to be. The revised code follows.

**Code:**

```
1   import numpy as np
2   import matplotlib.pyplot as plt
3   np.set_printoptions(precision=3,suppress=1,floatmode='fixed')
4   # define a piecewise function using if statements
5   # in this example, we have named the function pw
6   def pw(x):
7       #logic of the piecewise function
8       if x<=0:    #the condition is x<=0
9           y = x**2    #condition is true
10      else:       #otherwise
11          y = x+1 #condition is false
12      return y
13  vpw = np.vectorize(pw)
14  #set up the list for -2<x<=0
15  a = -2
16  b = 0
17  n = 50
18  dx = (b-a)/n
19  x = np.arange(a,b+dx,dx)    #last element may be very slightly above zero
20  x[n] = b                    #assign the last element to be zero
21  y = vpw(x)
22  plt.plot(x,y,'b')
23  #now do the second section of the function
24  a = 0
25  b = 2
26  n = 50
27  dx = (b-a)/n
28  # in this list we want to exclude the left endpoint at 0
29  # so we will use a start value that is slightly larger than 0
30  x = np.arange(a+dx,b+dx,dx)    #note, this includes 2
31  y = vpw(x)
32  plt.plot(x,y,'b')
33  plt.grid()
34  #plt.plot(0,0,'b.',markersize=11)
35  #plt.plot(0,1,'b.',fillstyle='none',markersize=11)
36  plt.show()
```

A new line 20 was inserted to force the right-hand endpoint to be $x = 0$. Now, the graph looks like this.

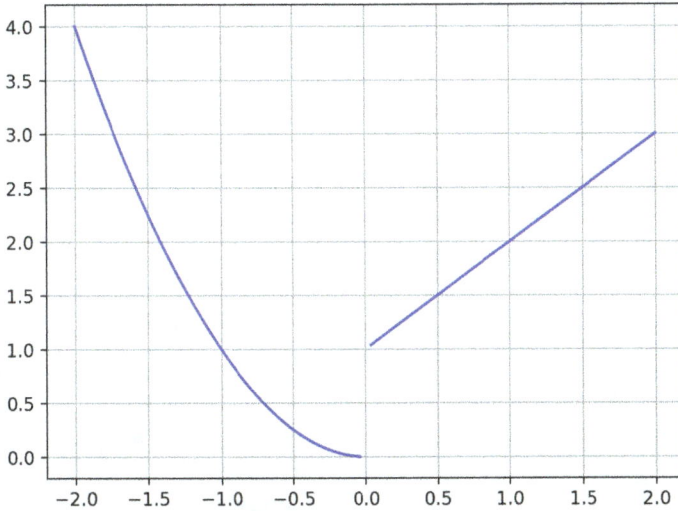

This graph is usually sufficient for informative and presentation purposes, but it does not explicitly show which part of the function is defined for $x = 0$. To make this clear, we can plot a closed dot on the part of the graph that is defined and an open dot where it is not. We use the subsequent lines to plot points at (0,0), which is defined by the function, and (0,1) which would be an open endpoint on the function.

```
plt.plot(0,0,'b.',markersize=11)
plt.plot(0,1,'b.',fillstyle='none',markersize=11)
```

The *markersize* parameter allows one to adjust the size of the point plotted. With these lines added, the graph then looks similar to those we would expect to see in a textbook.

Finally, we should point out that the *if* structure can have any number of pieces by using the *elif* option to add more conditional statements. For example, if we had a piecewise function with three parts, such as

$$f(x) = \begin{cases} x^2 & \text{if } x < -1 \\ x & \text{if } -1 \le x \le 1 \\ \sin x & \text{if } x > 1, \end{cases}$$

we could define the function as

```
def pw(x):
    if x<-1:
```

```
        y = x**2
    elif -1<=x<=1:
        y = x
    else:
        y = np.sin(x)
    return y
```

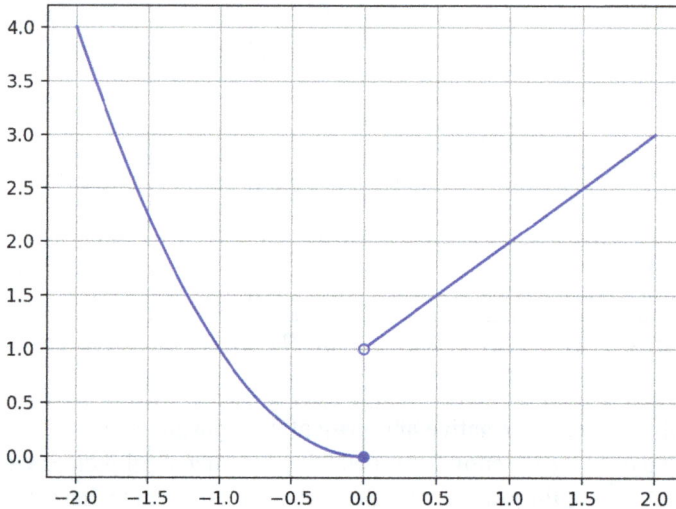

See Exercises 11–13.

## 3.6 Exercises

1. Use the print.format command and numpy package to modify the table in Figure 3.1 to include more angles and the tangent function as shown here:

```
angle |    0 |  π/6 |  π/4 |  π/3 |  π/2 | 2π/3 | 3π/4 | 5π/6 |    π |
---------------------------------------------------------------------
cos(x)|1.0000|0.8660|0.7071|0.5000|0.0000|-0.500|-0.707|-0.866|-1.000|
sin(x)|0.0000|0.5000|0.7071|0.8660|1.0000|0.8660|0.7071|0.5000|0.0000|
tan(x)|0.0000|0.5774|1.0000|1.7321| undef|-1.732|-1.000|-0.577|-0.000|
```

2. Write code that assigns 2 to a variable called *width* and $\sqrt{5}$ to a variable called *length*. Then, use a third variable called *area* to store the area of a rectangle with the specified length and width. Use print.format to display the result as follows: The area of a box with width xx.xx and length xx.xx is xx.xx, where the xx's are replaced with the appropriate values.

3. Write code to store the string 'Albert' in a variable called *firstname*. Then store the string 'Einstein' into a variable called *lastname*. Create a third variable called *fullname*. Use the variables and string operations to assign 'Einstein, Albert' to the variable *fullname*. Print the value of all three variables.

4. Assign the following string to a variable called *basetext*: 'Force is equal to the product of mass and acceleration.'
   (a) Search *basetext* for the substring 'mass'. Print the result.
   (b) Using the: notation, print just the word 'mass' from *basetext*.
   (c) Using the: notation, print the string beginning with the word 'product' through the end of the string.

5. Write code to accomplish the following tasks:
   (a) Define the function $height(t) = -16t^2 + 3t + 100$.
   (b) Evaluate the function at $t = 2$.
   (c) Output the following to the screen: The value of height at $t = 2$ is xx.xxx. (Use three decimal places in the format of the print statement.)

6. The *body mass index* is calculated according to the formula $\frac{weight}{height^2}$, where weight is in kg and height is in m. Write code to define a function called *bmi* that takes two arguments, height and weight, and returns the body mass index. Use the function to compute the body mass index of a person who is 1.7-m tall and weighs 68 kg. Print the result as follows: 'A person who is x.x m tall and weighs xx kg has a BMI of xx.xx.'

7. Write code to accomplish the following:
   (a) Prompt the user for their weight in kg.
   (b) Prompt the user for their height in m.
   (c) Convert the weight and height that were entered to numeric values.
   (d) Use the *bmi* function written in the previous problem to compute the BMI for the data entered.
   (e) Display the results in a meaningful message.

8. Modify the area example at the end of Section 3.4 so that the area is defined as a function with the radius as the argument of the function.

9. Use Python to plot the function $f(x) = \cos(x)$ between $x = 0$ and $x = 2\pi$.
   (a) Create a list containing the following x values: $0, \frac{\pi}{4}, \frac{\pi}{2}, \frac{3\pi}{4}, \pi, \frac{5\pi}{4}, \frac{3\pi}{2}, \frac{7\pi}{4}, 2\pi$.
   (b) Compute the corresponding y values for the given x values.
   (c) Use the format string 'go:' in the plot statement.

10. Add the graph of $g(x) = \sin(x)$ to the graph in Problem 9.
    (a) Draw $g(x)$ using a solid black line.
    (b) Include a legend that clearly identifies each function.
    (c) Include a grid in the plot.
    (d) Label the x axis with 'x'.

11. An object is thrown from the top of a 100-ft tall building. The height of the object above the ground after t seconds is given by $h(t) = -16t^2 + 10t + 100$.
    (a) At what time will the object hit the ground?

(b) Graph this function for $0 \le t \le b$, where $b$ is your answer to part (a). Use at least 50 points and include axis labels and a grid.

12. Suppose that a particular bacterial population grows exponentially for the first three hours and, then, because of environmental restrictions, the growth shifts to a rational function. Thus, the population of the bacteria is given by

$$P(t) = \begin{cases} e^t & \text{if } 0 \le t < 3 \\ \frac{52.5614t+3}{t+5} & \text{if } t \ge 3. \end{cases}$$

(a) Graph the population function on the interval $0 \le t \le 20$.
(b) Does it appear that the function is continuous?
(c) If we want the function to be smooth, what might we require?

13. Graph the function $f(x) = \begin{cases} x^2 & \text{if } x < -1 \\ x & \text{if } -1 \le x \le 1 \\ \sin x & \text{if } x > 1 \end{cases}$ on the interval $[-2, 3]$. Use at least 100 points. Include closed/open circles where appropriate.

# 4 Matrices, vectors, and linear systems

Our next topic deals with solving systems of linear equations. This should not be a new mathematical topic for students, but we wish to develop methods to solve these systems via programming. In the sciences, we often encounter large linear systems so employing computerized methods is a necessity. Before we begin the topic in earnest, we present some of the machinery and operations that apply to matrices and vectors in Python.

## 4.1 Matrices with numpy

A matrix is simply a rectangular array of numbers. The *shape* or *size* of a matrix is given by indicating the number of rows and columns contained in the matrix. For example, we say a matrix with three rows and four columns is 3 by 4 or $3 \times 4$. In mathematics, we enclose the array within square brackets. So, if

$$A = \begin{bmatrix} 1 & 2 & 3 & 4 \\ -2 & 4 & -3 & 5 \end{bmatrix},$$

then $A$ is a $2 \times 4$ matrix. A matrix with the same number of rows as columns is called *square*. The individual entries are denoted by row and column as $A(row, col)$. Hence, $A(2,4) = A_{2,4} = 5$. Sometimes, the corresponding lower case letters are used, and the comma notation is compressed. For example, we might write $A(2,4) = a_{24} = 5$. We will use the numpy library to do all of our matrix handling. The first step is to create a matrix with the desired numerical entries. In Python (and most programming languages) a matrix is simply a two-dimensional list (array)—a list of lists. If the matrix is relatively small, we can create the matrix explicitly with the desired values within the matrix. The following code shows one way to create a matrix, as well as some ways to access the entries of a matrix.

**Code:**

```
1  import numpy as np
2  #create the matrix
3  A = np.array([[1,2,3,4],[-2,4,-3,5],[-1,3,-3,4]])
4  print('A = ')
5  print(A)
6  #access the second row, third column
7  a23 = A[1,2]
8  print('The value in the second row, third column is ',a23)
9  #find the size of the matrix
10 m,n = np.shape(A)
11 print('rows = {}. cols = {}'.format(m,n))
```

https://doi.org/10.1515/9783110776645-004

```
12  #get the third row of the matrix
13  Arow3 = A[2,:]
14  print('The third row is ',Arow3)
15  #get the second column of the matrix
16  Acol2 = A[:,1]
17  print('The second column is ',Acol2)
```

**Output:**
```
A =
[[ 1  2  3  4]
 [-2  4 -3  5]
 [-1  3 -3  4]]
The value in the second row, third column is  -3
rows = 3. cols = 4
The third row is  [-1  3 -3  4]
The second column is  [2 4 3]

Process finished with exit code 0
```

The matrix is created in line 3 with the `.array` method. In line 7, we show how to access individual elements of the matrix, using the square brackets. The row is the first index, the column is the second index (remember, Python starts counting at 0). We can also use the ':' notation as we did with strings to access certain parts of the matrix. We can access all or part of any row or column. Lines 13 and 16 show how to access a certain row and column, respectively.

There are times when we may wish to retain the original matrix. Thus, we make a copy of the original, make changes to the copy, and then refer back to the original. We must be careful when we do this because matrices are said to be *immutable*. So using the '=' to make the copy may not act as you expect. See the following code and output.

**Code:**
```
1   import numpy as np
2   #create the matrix
3   A = np.array([[1,2,3,4],[-2,4,-3,5],[-1,3,-3,4]])
4   print('A = ')
5   print(A)
6   #show that matrices are immutable
7   B = A
8   #change B
9   B[0,0] = 2
10  #show that A was also changed.
```

```
11  print('A = ')
12  print(A)
```

**Output:**
```
A =
[[ 1  2  3  4]
 [-2  4 -3  5]
 [-1  3 -3  4]]
A =
[[ 2  2  3  4]
 [-2  4 -3  5]
 [-1  3 -3  4]]

Process finished with exit code 0
```

Note that $A[0,0]$ is originally set to be 1. The matrix $B$ is set equal to $A$, and then $B[0,0]$ is changed to be 2. However, when we print $A$ again, we see that $A[0,0]$ is now also 2. So changing $B$ also changed $A$. This is because, when we say that A = a matrix, it really means that $A$ points to a location in memory where the matrix is stored. So, when $B = A$, Python sets $B$ to point to same location in memory. Thus, if one of the variables is changed, then both of them are modified because they are both pointing to same location. We can accomplish what we need by asking for a 'hard' copy. We do so with the .copy method.

**Code:**
```
1   import numpy as np
2   #create the matrix
3   A = np.array([[1,2,3,4],[-2,4,-3,5],[-1,3,-3,4]])
4   print('Original A = ')
5   print(A)
6   #show that matrices are immutable
7   B = A.copy()
8   #change B
9   B[0,0] = 2
10  #show that A was also changed.
11  print('A after B has been changed =')
12  print(A)
13  print('B after the change =')
14  print(B)
```

**Output:**
```
Original A =
```

```
[[ 1  2  3  4]
 [-2  4 -3  5]
 [-1  3 -3  4]]
A after B has been changed =
[[ 1  2  3  4]
 [-2  4 -3  5]
 [-1  3 -3  4]]
B after the change =
[[ 2  2  3  4]
 [-2  4 -3  5]
 [-1  3 -3  4]]


Process finished with exit code 0
```

So, when we use A.copy and assign it to B, then we can change B without affecting A.

Let's investigate some of the other operators with respect to matrices. We want to see what +, −, *, and / do when we have matrices as the variables.

### 4.1.1 Addition and subtraction: $A \pm B$

We begin with addition and subtraction of $A$ and $B$.

**Code:**

```
1   import numpy as np
2   #create the matrix
3   A = np.array([[1,2],[3,4]])
4   B = np.array([[-1,3],[2,-5]])
5   print('A =')
6   print(A)
7   print('B =')
8   print(B)
9   #add two matrices
10  print('A+B = ')
11  print(A+B)
12  #subtract B from A
13  print('A-B = ')
14  print(A-B)
```

**Output:**

```
A =
[[1 2]
```

```
  [3 4]]
B =
[[-1  3]
 [ 2 -5]]
A+B =
[[ 0  5]
 [ 5 -1]]
A-B =
[[ 2 -1]
 [ 1  9]]

Process finished with exit code 0
```

$A+B$ creates a new matrix in which the corresponding elements of $A$ and $B$ are summed. Likewise $A-B$ subtracts each element of $B$ from the corresponding element in $A$. Because of the use of corresponding elements, $A$ and $B$ should be the same size in order to be able to add or subtract them. However, Python allows addition and subtraction of different-sized matrices in some circumstances. For example, the following code adds a 2×2 matrix to a $1 \times 2$ matrix.

**Code:**

```
1   import numpy as np
2   #create the matrix
3   A = np.array([[1,2],[3,4]])
4   B = np.array([[-1,3]])
5   print('A =')
6   print(A)
7   print('B =')
8   print(B)
9   #add two matrices
10  print('A+B = ')
11  print(A+B)
12  #subtract B from A
13  print('A-B = ')
14  print(A-B)
```

This produces the following output.

**Output:**

```
A =
[[1 2]
 [3 4]]
```

```
B =
[[-1  3]]
A+B =
[[0 5]
 [2 7]]
A-B =
[[ 2 -1]
 [ 4  1]]
```

```
Process finished with exit code 0
```

Because the number of columns in B was the same as that in A, the columns of B were added/subtracted to each of the rows of A. While we may find occasion to take advantage of this capability, it is generally bad practice, mathematically, to add or subtract matrices of different sizes.

### 4.1.2 Component-wise multiplication: $A * B$

We now investigate what happens when we use the multiplication symbol.

**Code:**

```
1   import numpy as np
2   #create the matrix
3   A = np.array([[1,2],[3,4]])
4   B = np.array([[-1,3],[2,-5]])
5   print('A =')
6   print(A)
7   print('B =')
8   print(B)
9   #component-wise multiplication
10  print('A*B = ')
11  print(A*B)
```

**Output:**

```
A =
[[1 2]
 [3 4]]
B =
[[-1  3]
 [ 2 -5]]
A*B =
```

```
[[ -1    6]
 [  6 -20]]
```

```
Process finished with exit code 0
```

For those that have already learned how to multiply matrices, it is clear that, in Python, $A * B$ does not yield the usual matrix product. We will discuss usual matrix multiplication later. The multiplication that is demonstrated here is called *component-wise* (or element-wise) multiplication. Each element of the resulting matrix is the product of the corresponding elements in $A$ and $B$, i. e., if $C = A * B$, the $C[i,j] = A[i,j] * B[i,j]$. Again, this should require that $A$ and $B$ be the same size, but Python allows the same kinds of scenarios that are allowed with addition and subtraction.

### 4.1.3 Component-wise division: $A/B$

If we change the "*" to "/" in the previous code, the following appears.

**Output:**
```
A =
[[1 2]
 [3 4]]
B =
[[-1   3]
 [ 2 -5]]
A/B =
[[-1.         0.66666667]
 [ 1.5       -0.8       ]]
```

```
Process finished with exit code 0
```

This demonstrates component-wise division, where $C_{ij} = A_{ij}/B_{ij}$.

As before, we see that Python defaults to displaying eight decimal places. This can make the display of larger matrices difficult to read. As we did in Section 3.5, we can set global printing options for matrices that restrict the number of decimals shown. The command has the following form:

```
np.set_printoptions(precision=3,suppress=1,floatmode='fixed')
```

Setting these global options modifies the display of the previous matrices to appear as follows.

**Output:**
```
A =
[[1 2]
 [3 4]]
B =
[[-1  3]
 [ 2 -5]]
A/B =
[[-1.000  0.667]
 [ 1.500 -0.800]]

Process finished with exit code 0
```

Notice that the final matrix entries display only three decimal places, but the first two matrices display integers. If any entry of the matrix is a floating-point number, then all entries are displayed with the designated precision.

### 4.1.4 Scalar multiplication: *cA*

We now consider multiplying a matrix by a number. Suppose we have a matrix $A$. If we multiply $A$ by a real number, $c$, then we simply multiply all entries of $A$ by $c$.

**Code:**
```
1   import numpy as np
2   np.set_printoptions(precision=3,suppress=1,floatmode='fixed')
3   #create the matrix
4   A = np.array([[1,2,3],[1,4,2],[2,-1,3]])
5   print('A =')
6   print(A)
7   #scalar multiplication
8   #multiply A by 3
9   print('3A = ')
10  print(3*A)
```

**Output:**
```
A =
[[ 1  2  3]
 [ 1  4  2]
 [ 2 -1  3]]
3A =
[[ 3  6  9]
```

```
[ 3 12  6]
[ 6 -3  9]]
```

```
Process finished with exit code 0
```

So, multiplying a matrix by a number, $c$, scales the matrix by a factor of $c$. Thus, we call $c$ a *scalar*, and we call this type of multiplication *scalar multiplication*.

### 4.1.5  Standard matrix multiplication

For those that have had linear algebra, the following section will not be new, but we need to spend some time developing a means of multiplying two matrices in such a way that the result is consistent with other mathematical principles. We are now familiar with component-wise multiplication, but mathematics provides a more meaningful definition of $AB$, provided that the two matrices are appropriately sized. To make our discussion more concise, we begin by defining some terms to be used later. A *row vector* is a matrix that has one row and a finite number of columns. If a matrix has three rows, then we could think of it as being composed of three row vectors. Likewise, a *column vector* is a matrix with a finite number of rows and one column. Thus, an example of a row vector is

$$\mathbf{v} = [\ 1 \quad 3 \quad -2 \quad 8 \quad -5\ ],$$

while a column vector would be something like

$$\mathbf{s} = \begin{bmatrix} 1 \\ 3 \\ -2 \\ -6 \\ 4 \end{bmatrix}.$$

In this example, $\mathbf{v}$ is $1 \times 5$, and $\mathbf{s}$ is $5 \times 1$. When the context is clear, we frequently drop the 'row' or 'column' designation and call either of them a vector. Further, when vectors are used, there is no need for the double-indexed notation. Hence, $v_1$ would denote first element of $\mathbf{v}$, $v_2$ the second element, and so forth. The *dot product* between two vectors is defined as follows:

Let $\mathbf{a} = [a_1\ a_2\ \dots\ a_n]$ and $\mathbf{b} = [b_1\ b_2\ \dots\ b_n]$. The dot product of $\mathbf{a}$ and $\mathbf{b}$ is

$$\mathbf{a} \cdot \mathbf{b} = a_1 b_1 + a_2 b_2 + \cdots + a_n b_n.$$

Written with summation notation, we have

$$\mathbf{a} \cdot \mathbf{b} = \sum_{i=1}^{n} a_i b_i.$$

Also, either of $\mathbf{a}$ or $\mathbf{b}$ could be a column vector, and the definition would not change.

**Example.** Let $\mathbf{a} = [1\ 3\ -2\ 8\ 5]$ and $\mathbf{b} = [1\ -2\ 4\ -6\ 4]$. Then

$$\mathbf{a} \cdot \mathbf{b} = (1)(1) + (3)(-2) + (-2)(4) + (8)(-6) + (5)(4)$$
$$= 1 + (-6) + (-8) + (-48) + (20)$$
$$= -41$$

With these concepts and terms, we can now define standard matrix multiplication. Let $A$ be an $m \times n$ matrix, and let $B$ be an $n \times k$ matrix. Then $AB$ is an $m \times k$ matrix such that $AB(r,c)$ is equal to the dot product of row $r$ of $A$ and column $c$ of $B$. That is,

$$A(r,c) = A_{r,:} \cdot B_{:,c}.$$

**Example.** Let $A = \left[\begin{smallmatrix} 1 & 2 & 3 \\ -1 & 2 & -3 \end{smallmatrix}\right]$ and $B = \left[\begin{smallmatrix} 1 & 2 \\ 0 & 4 \\ -3 & 2 \end{smallmatrix}\right]$. Then,

$$AB = \left[ \begin{array}{cc} 1(1) + 2(0) + 3(-3) & 1(2) + 2(4) + 3(2) \\ (-1)(1) + 2(0) + (-3)(-3) & (-1)(2) + 2(4) + (-3)(2) \end{array} \right]$$
$$= \left[ \begin{array}{cc} -8 & 16 \\ 8 & 0 \end{array} \right].$$

Let $A$ be $m \times n$ and $B$ be $n \times k$. Then, $AB$ has the following properties:
- The number of rows in $B$ must equal the number of columns in $A$. Otherwise the multiplication is not defined.
- The order of $A$ and $B$ matters, that is, $AB$ does not necessarily equal $BA$. In fact, sometimes one or the other is not defined.
- The matrix that results from the product $AB$ has as many rows as $A$ and as many columns as $B$.

Fortunately, Python has this multiplication already defined as the numpy.dot operator. The previous example is done in Python using the following code.

**Code:**
```
import numpy as np
np.set_printoptions(precision=3,suppress=1,floatmode='fixed')
#create the matrix
A = np.array([[1,2,3],[-1,2,-3]])
B = np.array([[1,2],[0,4],[-3,2]])
print('A =')
```

```
7   print(A)
8   print('B =')
9   print(B)
10  #standard matrix multiplication
11  C = np.dot(A,B)
12  print('AB =')
13  print(C)
```

**Output:**
```
A =
[[ 1  2  3]
 [-1  2 -3]]
B =
[[ 1  2]
 [ 0  4]
 [-3  2]]
AB =
[[-8 16]
 [ 8  0]]

Process finished with exit code 0
```

We see in line 11 that the .dot method requires that we specify the two matrices to be multiplied. The order of the matrices does matter because np.dot(A,B) performs *AB*, while np.dot(B,A) gives *BA*. In the code, the product is stored in a third matrix, C.

See Exercise 1.

## 4.2 Matrix inversion

In our usual real number system, we know that $(1)x = x$ for any value of $x$. Likewise, for any number $x \neq 0$, we know that $(\frac{1}{x})x = 1$. In more general (abstract) terms, we call the number 1 the *multiplicative identity* of the real numbers. Furthermore, we call $\frac{1}{x}$ the *multiplicative inverse* of $x$. The product of a number and its inverse is 1 (the identity). As we define the usual arithmetic operations for matrices, we would also like also to define a multiplicative identity and a multiplicative inverse for matrices. Doing so will allow us to construct a means for solving large linear systems via Python (or other programming languages).

### 4.2.1 The identity matrix

The *identity matrix of size n*, $I_n$, is a square matrix ($n \times n$) with 1's along the main diagonal and 0's everywhere else. So,

$$I_2 = \begin{bmatrix} 1 & 0 \\ 0 & 1 \end{bmatrix} \quad \text{while } I_5 = \begin{bmatrix} 1 & 0 & 0 & 0 & 0 \\ 0 & 1 & 0 & 0 & 0 \\ 0 & 0 & 1 & 0 & 0 \\ 0 & 0 & 0 & 1 & 0 \\ 0 & 0 & 0 & 0 & 1 \end{bmatrix}$$

Provided the sizes of the matrices allow for multiplication, then $AI = A$, and $IA = A$. A Python example that uses the $A$ matrix from the preceding section is given here.

**Code:**

```
1   import numpy as np
2   np.set_printoptions(precision=3,suppress=1,floatmode='fixed')
3   #create the matrix
4   A = np.array([[1,2,3],[-1,2,-3]])
5   I3 = np.array([[1,0,0],[0,1,0],[0,0,1]])
6   I2 = np.array([[1,0],[0,1]])
7   print('A =')
8   print(A)
9   print('I3 =')
10  print(I3)
11  print('I2 =')
12  print(I2)
13  #standard matrix multiplication
14  C = np.dot(A,I3)
15  print('(A)(I3) =')
16  print(C)
17  D = np.dot(I2,A)
18  print('(I3)(A) =')
19  print(D)
20  print('(A)(I2) =')
21  print(np.dot(A,I2))
```

**Output:**

```
A =
[[ 1  2  3]
 [-1  2 -3]]
I3 =
```

```
[[1 0 0]
 [0 1 0]
 [0 0 1]]
I2 =
[[1 0]
 [0 1]]
(A)(I3) =
[[ 1  2  3]
 [-1  2 -3]]
(I3)(A) =
[[ 1  2  3]
 [-1  2 -3]]
(A)(I2) =
Traceback (most recent call last):
  File "/Users/WillMiles/Desktop/_Courses/SciComp/SciCompBook/BookCode/
        basics.py", line 21, in <module>
    print(np.dot(A,I2))
  File "<__array_function__ internals>", line 5, in dot
ValueError: shapes (2,3) and (2,2) not aligned: 3 (dim 1) != 2 (dim 0)


Process finished with exit code 1
```

In the code, $I_3$ is constructed in line 5 and $I_2$ is constructed in line 6. The product $AI - 3$ is computed in line 14, and $I_2 A$ is done in line 17. Notice that, when we tried to multiply $AI_2$ (line 21), our program crashed, giving us an error message indicating that the dimensions of the matrices did not align.

The identity matrix acts like a 1 in real numbers. When we multiply a number by 1, we just get the original number. That is, $a \times 1 = a$. Likewise, when allowed, $AI = A$ in the space of matrices.

### 4.2.2 The inverse of a matrix

In the real numbers, if $ax = 1$, then we can solve for $x$ to get $x = \frac{1}{a}$. We call $\frac{1}{a}$ the multiplicative inverse of $a$. Similarly, we would like to find the multiplicative inverse of a matrix. In general, only square matrices have inverses. Also, not all square matrices have an inverse, but we are getting ahead of ourselves. Let $A$ be a matrix of size $n \times n$. We say that $X$ is the *inverse of A* if $AX = XA = I_n$. We write $X = A^{-1}$. Note that $A^{-1} \neq \frac{1}{A}$ since $\frac{1}{A}$ would not make sense in the space of matrices. The exponent notation is used only to denote that $A^{-1}$ is the multiplicative inverse of the matrix $A$ as just defined. Finding the inverse of a matrix requires the use of an algorithm to manipulate the rows of the matrix in such a way that the inverse is obtained. We will discuss some of this manipulation algorithm in the next section, but it is not our primary interest in this case. Thankfully, Python has a method that will find the inverse of a matrix. The method is located in the linear algebra routines which are included in numpy. An example is given next.

**Code:**

```
1   import numpy as np
2   #create the matrix
3   A = np.array([[1,2,3],[-1,2,-3],[0,2,5]])
4   print('A =')
5   print(A)
6   #find the inverse of A
7   A_inv = np.linalg.inv(A)
8   print('A_Inverse =')
9   print(A_inv)
10  #confirm the inverse
11  print('AA_inv = ')
12  print(np.dot(A,A_inv))
13  print('A_invA = ')
14  print(np.dot(A_inv,A))
```

**Output:**

```
A =
[[ 1  2  3]
 [-1  2 -3]
 [ 0  2  5]]
A_Inverse =
[[ 0.8  -0.2  -0.6 ]
 [ 0.25  0.25  0.  ]
 [-0.1  -0.1   0.2 ]]
AA_inv =
[[ 1.00000000e+00  2.77555756e-17 -5.55111512e-17]
 [-2.77555756e-17  1.00000000e+00  5.55111512e-17]
 [-2.77555756e-17 -2.77555756e-17  1.00000000e+00]]
A_invA =
[[ 1.0000000e+00  0.0000000e+00 -4.4408921e-16]
 [ 0.0000000e+00  1.0000000e+00  0.0000000e+00]
 [ 0.0000000e+00  0.0000000e+00  1.0000000e+00]]

Process finished with exit code 0
```

This program illustrates a few things about Python (and programming, in general). First, the inverse of the matrix $A$ is found in line 7 by using the numpy.linalg.inv(A) method. The inverse $A^{-1}$ is found to be

$$A^{-1} = \begin{bmatrix} 0.8 & -0.2 & -0.6 \\ 0.25 & 0.25 & 0 \\ -0.1 & -0.1 & 0.2 \end{bmatrix}.$$

Line 11 prints the result of $AA^{-1}$ which should be $I$. However, the numbers do not seem like ones and zeros. This is because of the way a computer performs arithmetic. It frequently must approximate numbers, and, hence, often accumulates rounding errors. We, the scientists and programmers, must recognize what is supposed to be a zero. For example, $A^{-1}(1,2) = 2.7755756 \times 10^{-17}$. Thus, this number has sixteen leading zeros in the decimal places. It is very very small, and we should recognize this number as the computer trying to say 0. To make the printout look nicer, we could use the print options command that we have used earlier, or we can round the entries of the matrix to a given number of decimal places. To accomplish the rounding, we use the command .round(*number of decimals*). The code to round the entries follows.

**Code:**

```
1   #confirm the inverse
2   print('AA_inv = ')
3   AA_inv = np.dot(A,A_inv)
4   #round the entries to 3 decimal places when printing
5   print(AA_inv.round(3))
6   print('A_invA = ')
7   A_invA=np.dot(A_inv,A)
8   print(A_invA.round(3))
```

**Output:**

```
AA_inv =
[[ 1.  0. -0.]
 [-0.  1.  0.]
 [-0. -0.  1.]]
A_invA =
[[ 1.  0. -0.]
 [ 0.  1.  0.]
 [ 0.  0.  1.]]

Process finished with exit code 0
```

Note that, when we use the .round method (in lines 5 and 8), it does not actually change the values that are stored in the matrix. If one wishes to use the rounded values, they must be stored in another variable. For example, we could use the command

D = AA_inv.round(3). This would stored the rounded values of AA_inv in the variable D.

If a matrix does not have an inverse, the method will fail, and an error is reported indicating that the matrix is *singular*.

**Code:**

```
1   import numpy as np
2   #np.set_printoptions(precision=3,suppress=1,floatmode='fixed')
3   #create the matrix
4   A = np.array([[1,2,3],[1,2,3],[0,2,5]])
5   print('A =')
6   print(A)
7   #find the inverse of A
8   A_inv = np.linalg.inv(A)
9   print('A_Inverse =')
10  print(A_inv)
```

**Output:**

```
Traceback (most recent call last):
  File "/Users/WillMiles/Desktop/_Courses/SciComp/SciCompBook/BookCode/
        basics.py", line 8, in <module>
    A_inv = np.linalg.inv(A)
  File "<__array_function__ internals>", line 5, in inv
  File "/Library/Frameworks/Python.framework/Versions/3.8/lib/python3.8/
        site-packages/numpy/linalg/linalg.py", line 545, in inv
    ainv = _umath_linalg.inv(a, signature=signature, extobj=extobj)
  File "/Library/Frameworks/Python.framework/Versions/3.8/lib/python3.8/
        site-packages/numpy/linalg/linalg.py",
        line 88, in _raise_linalgerror_singular
    raise LinAlgError("Singular matrix")
numpy.linalg.LinAlgError: Singular matrix
A =
[[1 2 3]
 [1 2 3]
 [0 2 5]]

Process finished with exit code 1
```

Our primary use of matrix inverses is to solve linear systems of equations, which we will discuss in the next section. However, readers especially interested in matrix algebra and inverses should consider taking a course in *linear algebra*.

**See Exercise 2.**

## 4.3 Linear systems

We now turn our attention to solving linear systems of equations. We studied these in algebra, but here we examine them from the perspective of using a computer and programs to obtain solutions in a faster, more efficient way. Let's start with an example. Consider the following set of equations:

$$x + y + z = 6 \tag{4.1}$$
$$2x - 3y + z = -1 \tag{4.2}$$
$$x + 2y - 3z = -4. \tag{4.3}$$

An equation that contains only numbers or variables to the first power, along with possible coefficients (no variables under square root symbols, no variables in denominators, no variables multiplied by each other), is called a *linear equation*. Each of the previous equations is linear. We call such a set of equations a *linear system of equations*. In most algebra courses, we learn to solve this type of system by using *elimination by addition*. The general method is to choose two of the equations, multiple them by constants in such a way that, when the modified equations are added, one of the variables is eliminated. Then, we choose another pair of equations and eliminate the same variable that was previously eliminated. For our current example, this process would generate two equations in two variables. We then multiply the new equations by constants so that adding will eliminate a variable. That will make it trivial to solve for the remaining variable. Finally, substitute the known value into one of the equations that has just two variables to find the value of a second variable. Finally, substitute both values into one of the original equations to find the value of the third variable. Let's work through the process.

We can choose any pair of equations. Say we choose equations (4.1) and (4.2). We can cause the coefficients of $x$ to be opposites of each other by multiplying (4.1) by $-2$ to get

$$-2x - 2y - 2z = -12$$
$$2x - 3y + z = -1.$$

Adding these two equations gives

$$-5y - z = -13.$$

Next, we'll choose equations (4.1) and (4.3). We can eliminate $x$ by multiplying one of the equations by $-1$. This gives

$$-x - y - z = -6$$
$$x + 2y - 3z = -4.$$

Adding these yields

$$y - 4z = -10.$$

Now, we have a system of two equations

$$-5y - z = -13$$
$$y - 4z = -10.$$

Next, we can eliminate $y$ by multiplying the second equation by 5 to get

$$-5y - z = -13$$
$$5y - 20z = -50.$$

Adding them gives

$$-21z = -63,$$

which implies that $z = 3$. Substitute $z = 3$ into $y - 4z = -10$ to get $y - 4(3) = -10$. Thus, $y = 2$. Finally, we substitute $z = 3$ and $y = 2$ into any equation that contains $x$. So $x + y + z = 6$ becomes $x + 2 + 3 = 6$, and $x = 1$. The process of substituting values back into the equations to find the values of other variables is called *back substitution.*

If we had a system of four equations with four unknowns, we would choose three pairs of equations and eliminate the same variable in each pair. This would yield a system of three equations in three unknowns. Then, we would proceed as in the previous example. So you can see how the process would work as the systems get larger. This repetitive procedure is exactly what lends itself to computation. However, programming languages don't really have the ability to deal with a function explicitly. Hence, we need to express the system in a different way. To do so, we will use matrices.

Let's return to our previous example where we wished to solve the system

$$x + y + z = 6$$
$$2x - 3y + z = -1$$
$$x + 2y - 3z = -4.$$

To create a matrix that represents the system, we take the coefficients of each variable from each equation and place them in a row. Likewise, we place the right-hand constant in the associated row. For example, the first row would be

$$1 \quad 1 \quad 1 \quad 6.$$

When we include the second and third row, we have

$$
\begin{array}{cccc}
1 & 1 & 1 & 6 \\
2 & -3 & 1 & -1 \\
1 & 2 & -3 & -4
\end{array}.
$$

Standard matrix notation places square brackets around the array of numbers. In the special case of representing equations, a vertical line is often placed in the matrix to separate the left sides of the equations from the right sides. Thus, we represent the system as follows:

$$
\left[
\begin{array}{ccc|c}
1 & 1 & 1 & 6 \\
2 & -3 & 1 & -1 \\
1 & 2 & -3 & -4
\end{array}
\right]. \tag{4.4}
$$

The matrix of coefficients of the variables along with the right-hand sides of the equations is called the *augmented matrix* for the system.

We have discussed how the computer can store and manipulate this type of structure. Since each row represents an equation, we can operate on the rows as we would an equation. That is, we are allowed to multiply or divide rows by nonzero numbers, and we can add two rows together and subtract rows from each other. We cannot, however, remove a row entirely. The matrix size must remain the same. Our goal is to find the values of $x, y$, and $z$. That is, we want

$$
x = n_1
$$
$$
y = n_2
$$
$$
z = n_3,
$$

where $n_1, n_2$, and $n_3$ are the values of the variables that solve the system. If we place this in matrix form, we have

$$
\left[
\begin{array}{ccc|c}
1 & 0 & 0 & n_1 \\
0 & 1 & 0 & n_2 \\
0 & 0 & 1 & n_3
\end{array}
\right]. \tag{4.5}
$$

Notice that the left block of the goal matrix is $I$. Thus, we wish to use "equation combinations" to get from (4.4) to (4.5). These equation combinations are called *row operations*, and the process that we follow is called *Gaussian elimination*. The order to be followed is important so that we avoid undoing progress made in a previous step while performing the next step in the process. In general, we will "fix" columns from left to right. We perform an operation that achieves a 1 in the appropriate location. Then, we perform operations to force the remainder of the column to contain zeros. Further, we get the ones by using multiplication and division, and we get the zeros by combining multiples of one equation with another. Sounds complicated, but an example will clear things up. So, we start with the matrix in (4.4)

$$\begin{bmatrix} 1 & 1 & 1 & | & 6 \\ 2 & -3 & 1 & | & -1 \\ 1 & 2 & -3 & | & -4 \end{bmatrix},$$

and we begin with the first column. Keeping the goal in mind, we wish to have a 1 in the first row. Fortunately, we already have that, so no work is required. Now, we wish to get zeros in rows two and three. Consider row two. We want a 0 in place of the 2. We are not allowed to simply subtract 2 from all the elements in the row. But, if we multiply the first row by –2 (which is allowed), we would get

$$\begin{bmatrix} -2 & -2 & -2 & | & -12 \\ 2 & -3 & 1 & | & -1 \\ 1 & 2 & -3 & | & -4 \end{bmatrix}.$$

The new equation represented by the first row is equivalent to the original equation (has the same solution). Now, adding rows one and two is equivalent to adding the two associated equations

$$\begin{array}{r} [-2 \quad -2 \quad -2 \quad -12] \\ + \;\; [\;\; 2 \quad -3 \quad\;\; 1 \quad\;\; -1] \\ \hline 0 \quad -5 \quad -1 \quad -13. \end{array}$$

We will put the result of the addition into row two, replacing the current row. This gives

$$\begin{bmatrix} -2 & -2 & -2 & | & -12 \\ 0 & -5 & -1 & | & -13 \\ 1 & 2 & -3 & | & -4 \end{bmatrix}.$$

We now have the zero where the 2 was, which is what we wanted. Now, we need to put the first row back the way it was so that we have the one that we wanted:

$$\begin{bmatrix} 1 & 1 & 1 & | & 6 \\ 0 & -5 & -1 & | & -13 \\ 1 & 2 & -3 & | & -4 \end{bmatrix}.$$

So, at this point, we have multiplied the first row by –2, added it to the second row, and replaced the second row. We will represent this operation like this:

$$-2R_1 + R_2 \rightarrow R_2.$$

Okay, now we need a zero in the third row, first column. If we call the matrix $A$, we can denote position in the matrix by $A(row, column)$. So, we want $A(3, 1)$ to be zero. We can accomplish this by subtracting the first row from the third row:

$$R_3 - R_1 \rightarrow R_3.$$

This gives

$$\left[\begin{array}{ccc|c} 1 & 1 & 1 & 6 \\ 0 & -5 & -1 & -13 \\ 0 & 1 & -4 & -10 \end{array}\right].$$

The first column is complete. Now move to the second column. According the goal matrix (4.5), we want $A(2,2) = 1$. To get this, we will divide the second row by $-5$

$$\frac{-1}{5}R_2 \rightarrow R_2.$$

This gives us

$$\left[\begin{array}{ccc|c} 1 & 1 & 1 & 6 \\ 0 & 1 & \frac{1}{5} & \frac{13}{5} \\ 0 & 1 & -4 & -10 \end{array}\right].$$

Now, get zeros in the other rows of the column:

$$R_1 - R_2 \rightarrow R_1$$
$$R_3 - R_2 \rightarrow R_3.$$

Then, we have

$$\left[\begin{array}{ccc|c} 1 & 0 & \frac{4}{5} & \frac{17}{5} \\ 0 & 1 & \frac{1}{5} & \frac{13}{5} \\ 0 & 0 & -\frac{21}{5} & -\frac{63}{5} \end{array}\right].$$

Now, we have two of the three columns accomplished. Next, we force $A(3,3) = 1$ with

$$\frac{-5}{21}R_3 \rightarrow R_3.$$

This gives

$$\left[\begin{array}{ccc|c} 1 & 0 & \frac{4}{5} & \frac{17}{5} \\ 0 & 1 & \frac{1}{5} & \frac{13}{5} \\ 0 & 0 & 1 & 3 \end{array}\right].$$

We get the necessary zeros with

$$\frac{-4}{5}R_3 + R_1 \rightarrow R_1$$
$$\frac{-1}{5}R_3 + R_2 \rightarrow R_2.$$

This gives

$$\left[ \begin{array}{ccc|c} 1 & 0 & 0 & 1 \\ 0 & 1 & 0 & 2 \\ 0 & 0 & 1 & 3 \end{array} \right].$$

And so we are finally done. We can now see the solution

$$x = 1, \quad y = 2, \quad z = 3.$$

It's a lot of work, especially by hand. So one can see that it would be beneficial to write a program (or use one that has already been written) to solve this. This would save a lot of time and produce far fewer errors. We can use our knowledge of matrices and the idea of row operations to solve the previous example in Python as follows.

**Code:**

```
1   import numpy as np
2   #np.set_printoptions(precision=3,suppress=1,floatmode='fixed')
3   #create the matrix
4   A = np.array([[1,1,1,6],[2,-3,1,-1],[1,2,-3,-4]])
5   print('A =')
6   print(A)
7   #perform row operations to achieve the goal matrix
8   #-2R1+R2-->R2
9   print('-2R1+R2-->R2')
10  A[1,:] = -2*A[0,:]+A[1,:]
11  print('A =')
12  print(A)
13  #-R1+R3-->R3
14  print('-R1+R3-->R3')
15  A[2,:] = -1*A[0,:]+A[2,:]
16  print('A =')
17  print(A)
18  print('(-1/5)R2-->R2')
19  A[1,:] = (-1/5.0)*A[1,:]
20  print('A =')
21  print(A)
22  print('-1R2+R1-->R1')
23  print('-1R2+R3-->R3')
24  A[0,:] = -1*A[1,:]+A[0,:]
25  A[2,:] = -1*A[1,:]+A[2,:]
26  print('A =')
```

```
27   print(A)
28   print('(-1/4)R3-->R3')
29   A[2,:] = (-1/4.0)*A[2,:]
30   print('A =')
31   print(A)
32   print('-1R3+R1-->R1')
33   A[0,:] = -1*A[2,:]+A[0,:]
34   print('A =')
35   print(A)
```

**Output:**
```
A =
[[ 1  1  1  6]
 [ 2 -3  1 -1]
 [ 1  2 -3 -4]]
-2R1+R2-->R2
A =
[[  1   1   1   6]
 [  0  -5  -1 -13]
 [  1   2  -3  -4]]
-R1+R3-->R3
A =
[[  1   1   1   6]
 [  0  -5  -1 -13]
 [  0   1  -4 -10]]
(-1/5)R2-->R2
A =
[[  1   1   1   6]
 [  0   1   0   2]
 [  0   1  -4 -10]]
-1R2+R1-->R1
-1R2+R3-->R3
A =
[[  1   0   1   4]
 [  0   1   0   2]
 [  0   0  -4 -12]]
(-1/4)R3-->R3
A =
[[1 0 1 4]
 [0 1 0 2]
 [0 0 1 3]]
-1R3+R1-->R1
```

```
A =
[[1 0 0 1]
 [0 1 0 2]
 [0 0 1 3]]
```

```
Process finished with exit code 0
```

The last matrix printed shows that we achieve the same solution as before: $x = 1$, $y = 2$, $z = 3$.

While we are less likely to make errors by performing the row operations in Python, it is still quite tedious to write the code to solve the system. Furthermore, if we had a large system, say 1,000 equations with 1,000 unknowns, then writing each row operation would be even more of a challenge. Thus, we need some way to automate our solution algorithm. There are many ways to accomplish this automation. We begin with expressing our system in yet another way. We return to our example system,

$$x + y + z = 6$$
$$2x - 3y + z = -1$$
$$x + 2y - 3z = -4$$

Let $A$ be the matrix of coefficients, without the right-hand sides. So,

$$A = \begin{bmatrix} 1 & 1 & 1 \\ 2 & -3 & 1 \\ 1 & 2 & -3 \end{bmatrix}.$$

Also, let $X$ be a column vector of our variables, and let $B$ be a column vector of the right-hand sides of the equations. Thus,

$$X = \begin{bmatrix} x \\ y \\ z \end{bmatrix}, \quad B = \begin{bmatrix} 6 \\ -1 \\ -4 \end{bmatrix}.$$

Now, consider $AX$.

$$AX = \begin{bmatrix} 1 & 1 & 1 \\ 2 & -3 & 1 \\ 1 & 2 & -3 \end{bmatrix} \begin{bmatrix} x \\ y \\ z \end{bmatrix}$$
$$= \begin{bmatrix} (1)x + (1)y + (1)z \\ 2x + (-3)y + (1)z \\ (1)x + 2y + (-3)z \end{bmatrix}$$

$$= \begin{bmatrix} x + y + z \\ 2x - 3y + z \\ x + 2y - 3z \end{bmatrix}$$

So, the entries of $AX$ are exactly the left-hand sides of the original equations. Since $B$ holds the right-hand sides, we can express the original system as $AX = B$. Now, we can use our knowledge of matrix inverses to help us solve the system. Multiplying both sides on the left by $A^{-1}$ gives

$$AX = B$$
$$A^{-1}AX = A^{-1}B$$
$$IX = A^{-1}B$$
$$X = A^{-1}B.$$

We can easily do this in Python.

**Code:**

```
1   import numpy as np
2   #create the coefficient matrix
3   A = np.array([[1,1,1],[2,-3,1],[1,2,-3]])
4   #create the right hand side column vector
5   B = np.array([[6],[-1],[-4]])
6   print('A =')
7   print(A)
8   print('B =')
9   print(B)
10  #get the inverse of A
11  AInv = np.linalg.inv(A)
12  #multiply the inverse of A by B
13  X = np.dot(AInv,B)
14  print('X =')
15  print(X)
```

**Output:**

```
A =
[[ 1  1  1]
 [ 2 -3  1]
 [ 1  2 -3]]
B =
[[ 6]
 [-1]
 [-4]]
```

```
X =
[[1.]
 [2.]
 [3.]]
```

```
Process finished with exit code 0
```

There is yet another way that Python can help to solve systems. The *linalg* methods includes a solve command. We still need to define the coefficient matrix and the right-hand side matrix. Then, we can find the solution with the *.linalg.solve* method.

**Code:**
```
1   import numpy as np
2   #np.set_printoptions(precision=3,suppress=1,floatmode='fixed')
3   #create the matrix
4   A = np.array([[1,1,1],[2,-3,1],[1,2,-3]])
5   B = np.array([[6],[-1],[-4]])
6   print('A =')
7   print(A)
8   print('B =')
9   print(B)
10  X = np.linalg.solve(A,B)
11  print('X =')
12  print(X)
```

**Output:**
```
A =
[[ 1  1  1]
 [ 2 -3  1]
 [ 1  2 -3]]
B =
[[ 6]
 [-1]
 [-4]]
X =
[[1.]
 [2.]
 [3.]]
```

```
Process finished with exit code 0
```

To close this discussion, we must recall from algebra that linear systems may have a unique solution (like the example problem we have been working with), an infinite number of solutions, or no solution. If the system has infinite solutions or no solution, we must recognize this as the practicing scientist. Python will report an error in either of the latter two cases. For example,

$$x + y + z = 6$$
$$2x + 2y + 2z = 12$$
$$x + 2y - 3z = -4$$

has an infinite number of solutions. But, if we try to solve it in Python, we get the following.

**Code:**

```
1  import numpy as np
2  #np.set_printoptions(precision=3,suppress=1,floatmode='fixed')
3  #create the matrix
4  A = np.array([[1,1,1],[2,2,2],[1,2,-3]])
5  B = np.array([[6],[12],[-4]])
6  print('A =')
7  print(A)
8  print('B =')
9  print(B)
10 X = np.linalg.solve(A,B)
11 print('X =')
12 print(X)
```

**Output:**

```
Traceback (most recent call last):
  File "/Users/WillMiles/Desktop/_Courses/SciComp/SciCompBook/BookCode/
        basics.py", line 10, in <module>
    X = np.linalg.solve(A,B)
  File "<__array_function__ internals>", line 5, in solve
  File "/Library/Frameworks/Python.framework/Versions/3.8/lib/python3.8/
        site-packages/numpy/linalg/linalg.py", line 393, in solve
    r = gufunc(a, b, signature=signature, extobj=extobj)
  File "/Library/Frameworks/Python.framework/Versions/3.8/lib/python3.8/
        site-packages/numpy/linalg/linalg.py",
            line 88, in _raise_linalgerror_singular
    raise LinAlgError("Singular matrix")
numpy.linalg.LinAlgError: Singular matrix
A =
```

```
[[ 1  1  1]
 [ 2  2  2]
 [ 1  2 -3]]
B =
[[ 6]
 [12]
 [-4]]

Process finished with exit code 1
```

On the other hand,

$$x + y + z = 6$$
$$2x + 2y + 2z = 6$$
$$x + 2y - 3z = -4$$

has no solutions, but Python will report the same error. We must be aware of these possibilities, and address them if they arise. We can address them via programming by examining something called the *determinant* of the matrix, but that is beyond the scope of this course. If this is of interest, we encourage the reader to consider taking a course in linear algebra.

As we stated earlier, one reason we wish to automate the solution of linear systems is to save time. For very large systems, entering the coefficients manually would take much time and tedious effort. Fortunately, when we encounter a very large system of equations, a couple of situations are common:

1. The coefficients associated with the equation have a particular pattern.
2. The coefficients are the result of other computations that are done in other portions of the problem-solving process.

In both of these scenarios, we can then automate the filling of the matrix entries with something called a *for loop*, which will be discussed in the next chapter.

**See Exercises 4–6.**

## 4.4 Exercises

1. Create the following matrices in Python using numpy.

$$A = \begin{bmatrix} 1 & -2 & 3 \\ 2 & 1 & 4 \\ 3 & -1 & -2 \end{bmatrix}, \quad B = \begin{bmatrix} 0 & 4 & 2 \\ 3 & -1 & -3 \end{bmatrix},$$

$$C = \begin{bmatrix} -2 & 1 \\ 0 & -1 \\ 1 & 3 \end{bmatrix}, \quad D = \begin{bmatrix} 1 & -3 & 0 \\ 2 & -2 & 2 \\ 3 & -1 & 1 \end{bmatrix}.$$

Use these matrices to display the results of the following operations or indicate that the operation is not defined.

(a) $A + D$
(b) $D - A$
(c) $3B$
(d) $A + B$
(e) $A * D$
(f) $A * C$
(g) $A/D$
(h) $A/B$
(i) $AD$, standard matrix multiplication
(j) $AB$, standard matrix multiplication
(k) $BC$, standard matrix multiplication

2. Find the inverse of matrix $A$ from exercise 1. Store the result in a variable named A_inv.

(a) Display A_inv using three decimal places for each entry.
(b) Display the result of $AA\_inv$ to show that $A^{-1}$ was computed correctly.

3. Indicate whether the following equations are linear.

(a) $2x - y = 7$
(b) $3x - 4 + 2z = 3y$
(c) $4\sqrt{x} - 2y = 10$
(d) $\frac{2}{x} - 4y + z = 6$
(e) $3xy - 2y + z = 4$
(f) $-x + 3y - \sin(z) = 2$

4. Solve the following system of equations by hand:

$$2x - y = 5$$
$$3x + 2y = 4.$$

5. Use matrix inverses or np.linalg.solve to solve the following system:

$$x - 3y + 2z - w = 6$$
$$2x - 4y + 5z + 2w = 13$$
$$-x + 3y - z + 3w = -23$$
$$3x + 2y - z - w = 6.$$

Print the solution as 'x = ??, y = ??, z = ??, w = ??'.

6. Solve the following system using Python and interpret the result:

$$x - 3y + 2z = 6$$
$$2x - 4y + 5z = 13$$
$$-x + 3y - 2z = -23.$$

# 5 Iteration

While solving linear systems is probably the most common task in applied mathematics, another often-used concept for solving problems algorithmically is iteration. We use iteration primarily for two functions:

- to advance in time;
- to improve upon a previous approximation.

To introduce iteration as a process, we begin by showing a powerful but simple method to find the roots of a function.

## 5.1 Finding roots: the bisection method

Suppose we have a continuous function $f(x)$, as shown below, and we wish to find the roots (or zeros) of the function, indicated by black dots. That is, we wish to solve

$$f(x) = 0.$$

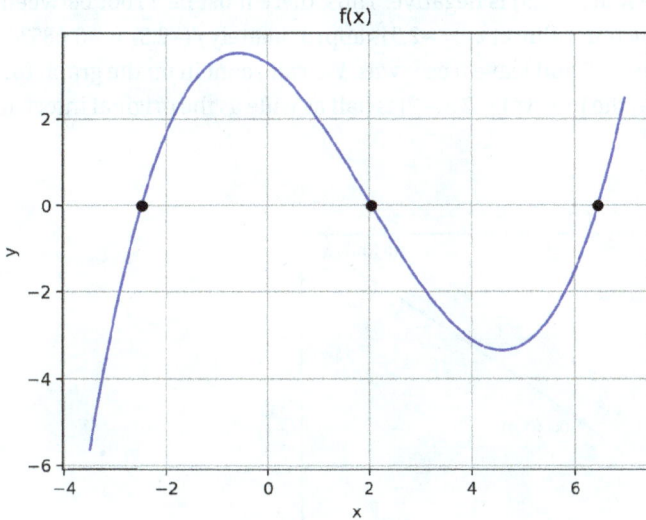

If we can find numbers $a$ and $b$ such that $f(a)$ and $f(b)$ have opposite signs, then the *intermediate value theorem* assures us that there is a value $c$, between $a$ and $b$, such that $f(c) = 0$. From the graph, we can see that $f(-3) < 0$ and $f(-2) > 0$. Therefore, there is a root between $x = -3$ (playing the role of $a$) and $x = -2$ (playing the role of $b$). Thus, we could take the midpoint between $a$ and $b$ as our first approximation of the root. Taking the midpoint is equivalent to bisecting the interval $[a, b]$. So, let $x_1 = \frac{a+b}{2}$. In our example, we have $x_1 = \frac{-3+-2}{2} = -2.5$ as can be seen in the next graph.

https://doi.org/10.1515/9783110776645-005

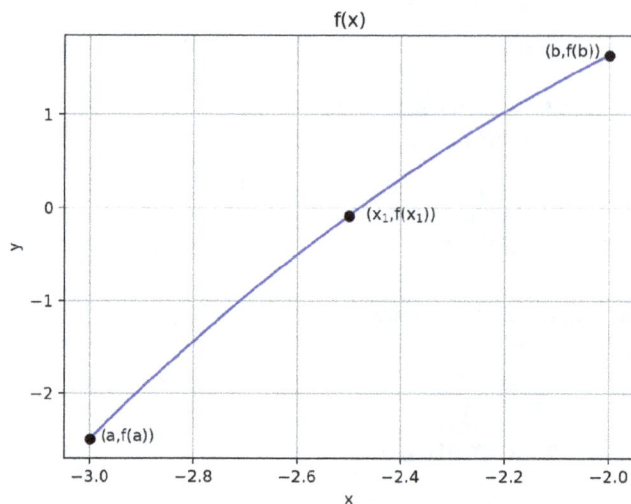

Now, if $f(x_1) = 0$, then we are done, and $x_1$ is a root. Otherwise, either: (a) $f(x_1)$ and $f(a)$ have different signs or (b) $f(x_1)$ and $f(b)$ have different signs. If (a) is true, then we know the root lies between $a$ and $x_1$. If (b) is true, then the root lies between $b$ and $x_1$. In the function shown, we see that $f(-2.5)$ is negative. Thus, there must be a root between $x = -2.5$ and $x = -2$. The function value at $x_1 = -2.5$ is approximately $f(-2.5) = -0.08534$. Now we can reassign $a$ to be $-2.5$ and leave $b$ as it was. We can zoom in on the graph for the new interval. Notice that the interval $[-2.5, -2]$ is half as wide as the original interval $[-3, -2]$.

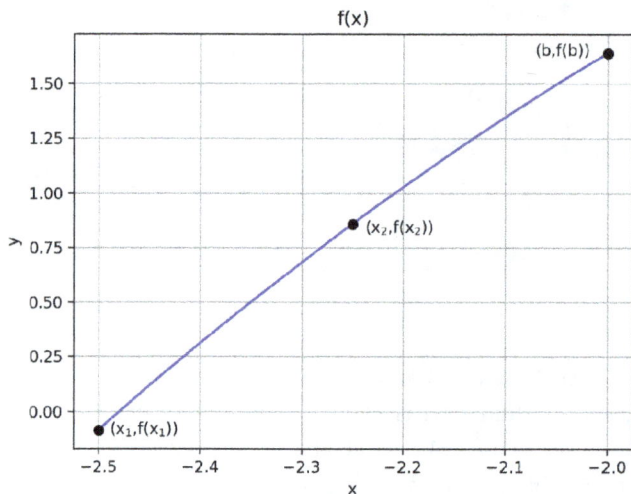

Now, $x_2 = \frac{-2.5 + -2}{2} = -2.25$. The value of the function at $x = -2.25$ is approximately $f(-2.25) = 0.85844$. This is actually further from 0 than our original guess, but, because the interval is shrinking, we can continue the process to get as close to the actual root as desired. We say that $x_1$ is the first approximation or the first *iterate*, $x_2$ is the second approximation or second *iterate*, and so on. From the graph, we see that $f(x_2)$ differs in sign from $f(-2.5)$. Thus, we can reassign $b$ to be $x_2 = -2.25$, leave $a$ as it was, and repeat the process.

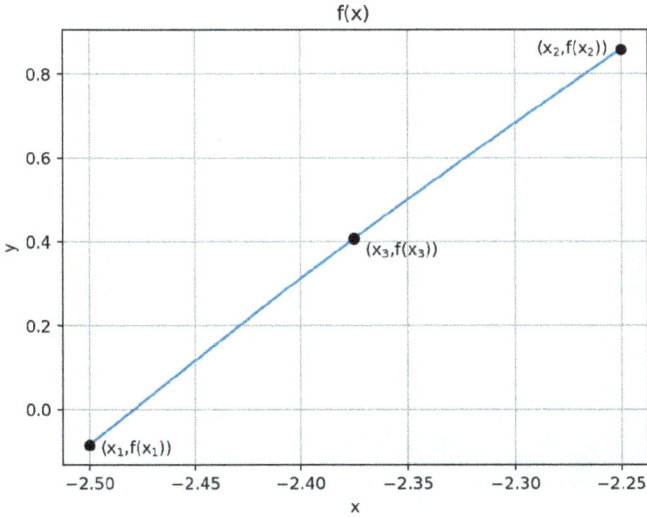

We can then find the midpoint of the new interval as the next approximation and continue the process until we reach an approximation that is either the exact root or "close enough." The sequence of the first five iterates, plotted on the $x$-axis, is shown in the next graph.

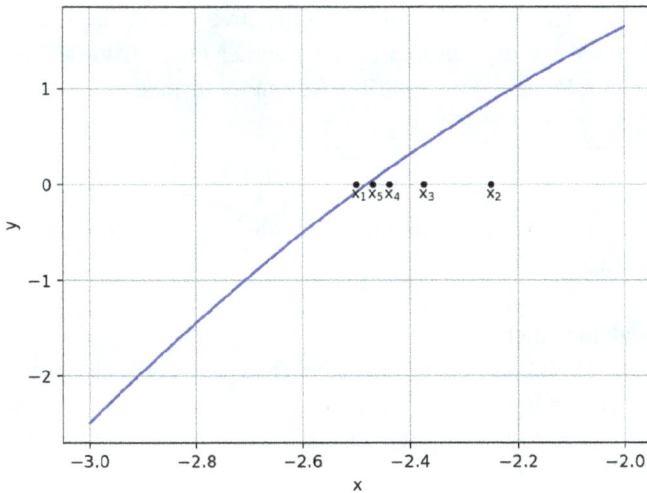

Note that the iterates would change if the starting interval were different from $[-3, -2]$. For example, if the starting interval was $[-2.75, -2]$, then the iterates would "bounce" around more because the root is closer to the center of the original interval. The sequence of iterates that corresponds to a starting interval of $[-2.75, -2]$ is given in the following graph.

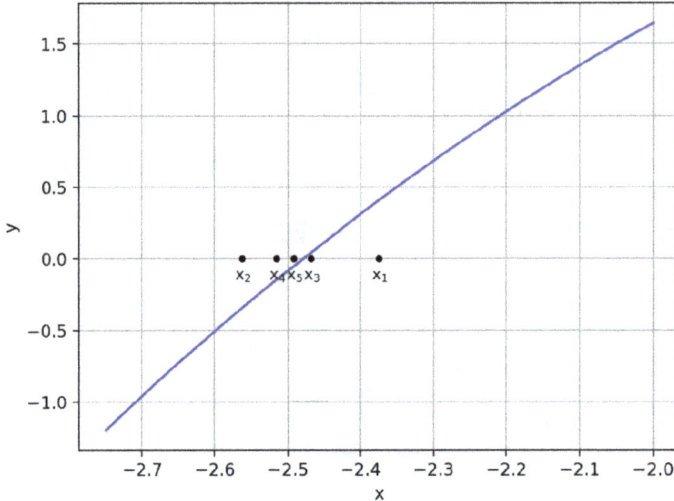

Because the process repeats itself, it is a good candidate for computation. Thus, the task becomes one of translating the process just described into an algorithm that we can program using Python. For any continuous function $f(x)$, we can use a graph to find original values for $a$ and $b$ so that $f(a)$ and $f(b)$ have different signs. From a programming standpoint, we can test whether two numbers, $c$ and $d$, have different signs by considering their product. If $cd$ is negative, then one of the numbers is negative and one is positive (so they have different signs). We can use this fact to test whether the function values have the same or different signs. We can now describe, with more formality, the *bisection algorithm*.

## Bisection Algorithm

Assume that $f(x)$ is a continuous function on the interval $[a, b]$ and that $f(a)$ and $f(b)$ have different signs, that is to say $f(a)f(b) < 0$.

1. Let $x = \frac{a+b}{2}$.
2. If $f(x) = 0$ or close enough to 0, then quit.
3. If $f(a)f(x) < 0$, then $f(a)$ and $f(x)$ have different signs. Thus, the root is between $a$ and $x$. Reassign $b$ to be $x$ and go back to step (1).
4. If $f(x)f(b) < 0$, $f(x)$ and $f(b)$ have different signs. Thus, the root is between $x$ and $b$. Reassign $a$ to be $x$ go back to (1).

Let's work through an example.

**Example.** Let $f(x) = \frac{x^4}{10} - 2x^2 - x - 3\sin(x) + 5$. Using Python to graph the function between $x = -5$ and $x = 5$ yields the following.

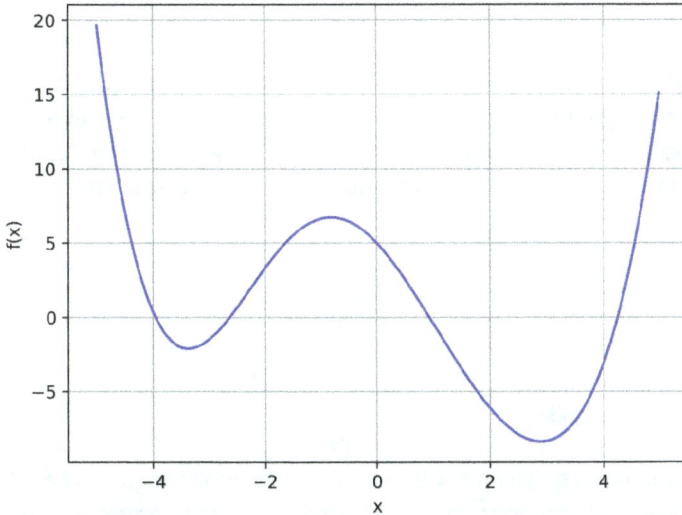

From the graph, we can see that $f(x)$ has multiple zeros. There is one that is very close to $x = -4$, one between $x = -4$ and $x = -2$, one between $x = 0$ and $x = 2$, and one between $x = 4$ and $x = 5$. There may be others zeros, but we can see those four on the graph presented. To illustrate the method, let's try to find the zero that is between $x = 0$ and $x = 2$. Note that $f(0) > 0$ and $f(2) < 0$. Thus, our first approximation to the root is $x = \frac{a+b}{2} = \frac{0+2}{2} = 1$. The values for $f(0), f(2)$, and $f(x_1)$ are found in Python as follows.

**Code:**

```
1   import numpy as np
2
3   def f(x):
4       y = x**4/10 -2*x**2 + -x-3*np.sin(x) + 5
5       return(y)
6
7   a = 0
8   b = 2
9   print('f({:.4f}) = {:.4f}'.format(a,f(a)))
10  print('f({:.4f}) = {:.4f}'.format(b,f(b)))
11  x = (a+b)/2.0
12  print('f({:.4f}) = {:.4f}'.format(x,f(x)))
```

**Output:**

```
f(0.0000) = 5.0000
f(2.0000) = -6.1279
f(1.0000) = -0.4244

Process finished with exit code 0
```

We can see that $f(0)$ and $f(1)$ have different signs. Thus, we now know that the root is between 0 and 1. So we reassign the right-hand side of our interval from $b = 2$ to $b = 1$. Then, we repeat the procedure. So, we can update our approximation to the root as the midpoint between 0 and 1, or $x = \frac{0+1}{2} = \frac{1}{2}$. When we do that, we get the following.

**Output:**

```
f(0.0000) = 5.0000
f(1.0000) = -0.4244
f(0.5000) = 2.5680

Process finished with exit code 0
```

Now, we see that the root must be between $x = 0.5$ and $x = 1$. So, we reassign $a$ to be 0.5. Each computational process that generates a new approximation (an iterate) is called an *iteration*. The next table summarizes the results of the first ten iterations.

| Iteration | a | b | x | f(x) |
|---|---|---|---|---|
| 1 | 0 | 2 | 1 | −0.4244 |
| 2 | 0 | 1 | 0.5 | 2.5680 |
| 3 | 0.5 | 1 | 0.75 | 1.1117 |
| 4 | 0.75 | 1 | 0.875 | 0.3497 |
| 5 | 0.875 | 1 | 0.9375 | −.0363 |
| 6 | 0.875 | 0.9375 | 0.9062 | 0.1570 |
| 7 | 0.9062 | 0.9375 | 0.9219 | 0.606 |
| 8 | 0.9219 | 0.9375 | 0.9297 | 0.0120 |
| 9 | 0.9297 | 0.9375 | 0.9336 | −0.0121 |
| 10 | 0.9297 | 0.9336 | 0.9316 | −0.0001 |

Note that, unless we wish to save the values of the iterates, we can simply replace the value of previous iterate with the current iterate. Thus, we can use the same variable (in this case, $x$) for all iterates instead of $x_1, x_2, \dots$. Manually changing $a$ and $b$ in the code for each iteration is unreasonable. To make the code loop back to perform another iteration, we will use a *while loop*. The general format of a while loop is

```
while condition:
    statements to be executed
```

The indentation is important. All the code that is indented under the 'while' statement is part of the loop. Some part of the indented code will need to update the variables that are involved in the while condition. The loop will be executed multiple times until the while condition is violated, at which point, the code will drop to the next line below the loop.

Since we are seeking a root of a function, we want to continue to generate iterates until $f(x)$ is very close to zero. Thus, we need something similar to the following.

```
while |f(x)| > 0.0001:
    statements to be executed
```

The "cutoff" value for our loop is called the *tolerance*. In the current example, the tolerance is 0.0001. We accomplish the reassignment of $a$ and $b$ by using the if statement that we learned earlier. We now present a version of the bisection algorithm for finding roots of a continuous function. The code includes comment that indicate each step of the algorithm.

**Code:**

```
1   import numpy as np
2
3   def f(x):
4       y = x**4/10 -2*x**2 + -x-3*np.sin(x) + 5
5       return(y)
6
7   tol = 0.0001
8   a = 0
9   b = 2
10  #we will do the first iterate before our while loop starts so that we
11  #have a value to test against the tolerance
12  x = (a+b)/2
13  while np.abs(f(x))>tol:
14      print('a={:.5f}  f(a)={:.5f},    b={:.5f}  f(b)={:.5f},   \
15          x={:.5f}  f(x)={:.5f}'.format(a,f(a),b,f(b),x,f(x)))
16      #now decide whether we replace a or b with x
17      if f(a)*f(x) < 0:
18          #root is between a and x so replace b
19          b = x
20      elif f(b)*f(x)<0:
21          #root is between b and x so replace a
22          a = x
23      else:
24          # in this case, f(x) must be 0 and we have found the root
```

```
25          # so we will know the root value is x and we can end the loop
26          break
27      #recompute the approximation
28      x = (a+b)/2
29  print('final x =',x)
30  print('final f(x) =',f(x))
```

**Output:**

```
a=0.00000  f(a)=5.00000,   b=2.00000  f(b)=-6.12789,  x=1.00000  f(x)=-0.42441
a=0.00000  f(a)=5.00000,   b=1.00000  f(b)=-0.42441,  x=0.50000  f(x)=2.56797
a=0.50000  f(a)=2.56797,   b=1.00000  f(b)=-0.42441,  x=0.75000  f(x)=1.11172
a=0.75000  f(a)=1.11172,   b=1.00000  f(b)=-0.42441,  x=0.87500  f(x)=0.34974
a=0.87500  f(a)=0.34974,   b=1.00000  f(b)=-0.42441,  x=0.93750  f(x)=-0.03631
a=0.87500  f(a)=0.34974,   b=0.93750  f(b)=-0.03631,  x=0.90625  f(x)=0.15703
a=0.90625  f(a)=0.15703,   b=0.93750  f(b)=-0.03631,  x=0.92188  f(x)=0.06043
a=0.92188  f(a)=0.06043,   b=0.93750  f(b)=-0.03631,  x=0.92969  f(x)=0.01208
a=0.92969  f(a)=0.01208,   b=0.93750  f(b)=-0.03631,  x=0.93359  f(x)=-0.01211
final x = 0.931640625
final f(x) = -1.3723513678343124e-05


Process finished with exit code 0
```

Notice that the absolute value is accomplished with the *np.abs* method. It is used in line 13 to see if the function value is close to zero.

Now that we have a working program for the bisection method, we would like to package it as a function so that we can use it to find roots for any appropriate function. To do this, we will define a function called bisect which takes four arguments: the name of the function for which to find the root, *a* (lower bound for the root), *b* (upper bound for the root), and tolerance. For the most part, we can simply wrap our existing code in the def structure, but there are a few important changes. See the following.

**Code:**

```
1   def bisect(f,a,b,tol):
2       #we will do the first iterate before our while loop starts so that we
3       #have a value to test against the tolerance
4       x = (a+b)/2
5       while np.abs(f(x))>tol:
6           print('a={:.5f}  f(a)={:.5f},    b={:.5f}  f(b)={:.5f}, \
7              x={:.5f}   f(x)={:.5f}'.format(a,f(a),b,f(b),x,f(x)))
8           #now decide whether we replace a or b with x
9           if f(a)*f(x) < 0:
10              #root is between a and x so replace b
11              b = x
12          elif f(b)*f(x)<0:
```

```
13          #root is between b and x so replace a
14          a = x
15      else:
16          # in this case, f(x) must be 0 and we have found the root
17          # so we will know the root value is x and we can end the loop
18          break
19      #recompute the approximation
20      x = (a+b)/2
21  return x
```

The first thing to notice about this code is the addition of a return statement in line 21. We must return the root that was found by the method. Also, the function f is now a local variable for the defined function bisect. In the bisect function, f is the variable that holds the name of the function whose root is sought. Thus, we can define any function for which we want to find a root and pass on the name to the bisection function. For example, suppose we wish to find the root of $f(x) = e^x - 3$. By inspection, we see that $f(1)$ is negative and $f(2)$ is positive. So, there is a root between $a = 1$ and $b = 2$. We could call the bisect function with the following few lines of code added below the function definition.

**Code:**

```
22  def shifted_exp(x):
23      y = np.exp(x) - 3
24      return y
25
26  tol = 0.0001
27  a = 1
28  b = 2
29  x = bisect(shifted_exp,a,b,tol)
30  print('final x =',x)
31  print('final f(x) =',shifted_exp(x))
```

**Output:**

```
a=0.00000  f(a)=-2.00000,  b=2.00000  f(b)=4.38906,  x=1.00000  f(x)=-0.28172
a=1.00000  f(a)=-0.28172,  b=2.00000  f(b)=4.38906,  x=1.50000  f(x)=1.48169
a=1.00000  f(a)=-0.28172,  b=1.50000  f(b)=1.48169,  x=1.25000  f(x)=0.49034
a=1.00000  f(a)=-0.28172,  b=1.25000  f(b)=0.49034,  x=1.12500  f(x)=0.08022
a=1.00000  f(a)=-0.28172,  b=1.12500  f(b)=0.08022,  x=1.06250  f(x)=-0.10640
a=1.06250  f(a)=-0.10640,  b=1.12500  f(b)=0.08022,  x=1.09375  f(x)=-0.01455
a=1.09375  f(a)=-0.01455,  b=1.12500  f(b)=0.08022,  x=1.10938  f(x)=0.03246
a=1.09375  f(a)=-0.01455,  b=1.10938  f(b)=0.03246,  x=1.10156  f(x)=0.00886
a=1.09375  f(a)=-0.01455,  b=1.10156  f(b)=0.00886,  x=1.09766  f(x)=-0.00287
a=1.09766  f(a)=-0.00287,  b=1.10156  f(b)=0.00886,  x=1.09961  f(x)=0.00299
```

```
final x = 1.0986328125
final f(x) = 6.15721275165626e-05

Process finished with exit code 0
```

The function is defined as shifted_exp (shifted exponential). Then, the root is found by calling the bisect function in line 8. Finally, once we are satisfied that the function is working properly, we should remove most of the print statements within the bisect function. Also, when the function is complete, we can save the function in a separate file and import the file when needed. For example, if we save the function from the def statement to the return statement (and include any lines that import packages) in a file called *bisectfun.py*, then we could accomplish the previous example with the following code.

**Code:**

```
1   import numpy as np
2   from bisectfun import *
3
4   def shifted_exp(x):
5       y = np.exp(x) - 3
6       return y
7
8   tol = 0.0001
9   a = 1
10  b = 2
11  x = bisect(shifted_exp,a,b,tol)
12  print('final x =',x)
13  print('final f(x) =',shifted_exp(x))
```

**See Exercises 1–5.**

## 5.2 Euler's method for differential equations

Another reason for iteration is to get approximations at different times for a given equation. This need occurs frequently when solving *differential equations*. A *differential equation* (DE) is an equation that contains derivatives. For example, $\frac{d^2y}{dx^2} + y = 3x - 4$ is a differential equation. It is a *second-order* equation because the highest derivative that appears is the second derivative. The *solution* to the DE is a function $y(t)$ that, when substituted into the equation, makes the equation true. Some differential equations are fairly easy to solve. If we have

$$\frac{dy}{dx} = 2x + 1,$$

then we could integrate both sides with respect to $x$ to get the solution:

$$\frac{dy}{dx} = 2x + 1$$

$$\int \frac{dy}{dx}\, dx = \int (2x + 1)\, dx$$

$$y(x) = \frac{2x^2}{2} + x + C$$

$$y(x) = x^2 + x + C,$$

where $C$ is the constant of integration. So, in this case, there are an infinite number of functions that satisfy the equation. We can check the result

$$\frac{d}{dx}(x^2 + x + C) =? \; 2x + 1$$

$$2x + 1 + 0 =? \; 2x + 1$$

$$2x + 1 = 2x + 1 \quad \checkmark$$

Other differential equations are very difficult (or impossible) to solve explicitly. We know that $\frac{dy}{dt} = e^{t^2}$ must have a solution, but it has also been shown that there is no closed-form solution for $\int e^{t^2}\, dt$. Thus, the best we can do is to try to get values of the desired function for a sequence of $t$ values. If we think of $t$ as time, then we are seeking approximations of $y$ for a specified set of times. One of the earliest and most straight-forward approaches to find such approximations is credited to Euler. *Euler's method* applies to first-order differential equations that can be arranged such that

$$\frac{dy}{dt} = f(y, t).$$

Some examples are

$$\frac{dy}{dt} = \frac{y - t}{2}$$

$$\frac{dy}{dt} = e^t - \cos(t)$$

$$\frac{dy}{dt} = e^{t^2}.$$

Recall from Calculus I that the slope of the tangent line to a function $g(t)$ at $t = a$ is given by $\frac{dg}{dt}(a) = g'(a)$, and the point of tangency is $(a, g(a))$. Then, we can use the point–slope form of a line to find the equation of the tangent line:

$$y - y_1 = m(t - t_1)$$

$$y - g(a) = g'(a)(t - a)$$

$$y = g'(a)(t - a) + g(a).$$

The critical idea in this method is that, when $t$ is near $a$, the tangent line is close to the function. Thus, the tangent line should provide a reasonable approximation to the function. So, when $t$ is near $a$, $g(t) \approx g'(a)(t - a) + g(a)$. For the advanced reader, the method depends on the Taylor series expansion of $g(t)$, but that knowledge is not assumed here. Now consider

$$\frac{dy}{dt} = \frac{y - t}{2}.$$

While we do not know the formula for the function $y(t)$, if we know a point on the function $(t_0, y_0)$, we can find the derivative of the function using this equation. Suppose we knew that, when $t = 0$, $y = 1$, and we want to approximate the function $y$ for $t = 0.5, 1.0, 1.5, 2.0, 2.5, \ldots, 5.0$. Note that the $t$ values are equally spaced. In this case, the successive values are 0.5 apart. We denote the spacing between $t$ values by $\Delta t$, substituting $t = 0$ and $y = 1$ into the differential equation

$$y'(0) = \frac{1 - 0}{2} = \frac{1}{2}.$$

Furthermore, the tangent line at $(0, 1)$ is given by

$$y_{\text{tan}} = y'(0)(t - 0) + y(0)$$
$$y_{\text{tan}} = \frac{1}{2}(t) + 1.$$

This means that, for $t$ near 0,

$$y(t) \approx \frac{1}{2}(t) + 1.$$

The closer $t$ is to 0, the better the approximation. Now we consider $t = 0.5$. Then,

$$y(0.5) \approx \frac{1}{2}(0.5) + 1 = 1.25.$$

Thus, we have an approximation for $y$ when $t = 0.5$. So, a point on the graph of $y$ is approximately $(0.5, 1.25)$. Since the slope of $y$ is not constant (how do we know this?), we update the slope and compute a new tangent line. We compute the slope of the tangent line using the approximation $(0.5, 1.25)$. Hence,

$$\frac{dy}{dt} = \frac{y - t}{2}$$
$$= \frac{1.25 - 0.5}{2}$$
$$= \frac{0.75}{2}$$
$$= 0.375$$

Then, the updated tangent line is given by

$$y - 1.25 = 0.375(t - 0.5)$$
$$y = 0.375(t - 0.5) + 1.25.$$

This allows us to obtain an approximation associated with our next $t$ value, $t = 1$.

$$y(1) \approx = y'(0.5)(1 - 0.5) + y(0.5)$$
$$= 0.375(0.5) + 1.25$$
$$= 1.4375.$$

Now, we have three points that approximate the function to be found: $(0, 1)$, $(0.5, 1.25)$, and $(1, 1.4375)$. A table of values could be made as follows.

| Iteration # | $t$ | $y$ |
|---|---|---|
| 0 | 0 | 1 |
| 1 | 0.5 | 1.25 |
| 2 | 1.0 | 1.4375 |

Notice that the $t$ values are evenly spaced by $\Delta t$. We will use subscripts to denote the iteration. So, $t_0 = 0, y_0 = 1, t_1 = 0.5, y_1 = 1.25$, and $t_2 = 1.0, y_2 = 1.4375$.

The general idea is to use the most recent approximation to compute the slope. Then, update the tangent line to find the next approximation. So, in general, we have a current approximation $(t_{current}, y_{current})$. Thus, the slope of the updated tangent line is given by $y'(t_{current}, y_{current})$, and the tangent line is found using the point–slope form of a line.

$$y - y_{current} = y'(t_{current}, y_{current})(t - t_{current})$$
$$y = y'(t_{current}, y_{current})(t - t_{current}) + y_{current}$$

Using this tangent line, we can approximate the next value of $y$ by

$$y_{next} = y'(t_{current}, y_{current})(t_{next} - t_{current}) + y_{current}.$$

If we denote the current iteration by $n$ and the next iteration by $n + 1$, then the previous equation can be written as

$$y_{n+1} = y'(t_n, y_n)(t_{n+1} - t_n) + y_n.$$

Since the $t$ values are evenly spaced, $t_{n+1} - t_n = \Delta t$ and the equation can be expressed as

$$y_{n+1} = y'(t_n, y_n)\Delta t + y_n.$$

This formula is called a *recursive* formula because the next term depends on the previous term(s) of the sequence. When we see recursion in mathematics, it often translates to a loop structure within the associated programming code. Thus, it should be possible to write a Python program to compute the Euler iterates for this differential equation. We wish to solve

$$\frac{dy}{dt} = \frac{y - t}{2}, \quad 0 \le t \le 5,$$

where $y(0) = 1$. The given point, $(0, 1)$, is called an *initial condition*, and it acts as the initial point in our set of approximation points. We are allowed to choose the spacing of the approximation points (or equivalently, the number of approximation points to compute). For any interval of real numbers $[a, b]$, if we have $n + 1$ points spaced uniformly through the interval, then we have $n$ subintervals. This means that the width of each subinterval ($\Delta t$) is given by

$$\Delta t = \frac{b - a}{n}.$$

So, let's put this together in Python. We begin by defining the right-hand side of the equation as a function in lines 7–9 as shown.

**Code:**

```
1   import numpy
2   import matplotlib.pyplot as plt
3
4   #solve dy/dt = (y-t)/2
5   #where when t=0, y=1
6
7   def rhs(t,y):
8       m = (y-t)/2
9       return m
```

We let $a$ be the initial $t$ value and $b$ be the final $t$ value. Suppose we will to have eleven iterates, then $n = 10$. This allows us to compute $\Delta t$ as follows.

**Code:**

```
10   #initial t value
11   a = 0
12   #final t value
13   b = 5
14   #number of intervals
15   n = 10
```

```
16   #delta t
17   dt = (b-a)/n
```

We know that, for each iterate, $t$ will have to increase by an amount of $\Delta t$. The starting value of $t$ is 0, and the ending value of $t$ is 5. There are many ways to implement this requirement. One way is to use the *.arange* method as shown here:

```
t = np.arange(a,b+dt,dt)
```

This produces a list of the necessary $t$ values. Now, we need to compute the approximation values that are associated with each $t$ value. We will use the formula for $y_{n+1}$ to do this, but we also need something called a *for loop*. We pause our discussion of Euler's method briefly to explain how such a loop will work.

A for loop is similar to a while loop except that it loops over a particular set of *index* values. For example, consider what follows.

**Code:**
```
1   import numpy as np
2   t = np.arange(0,10,1)
3   print(t)
4   for i in t:
5       print(i)
```

Line 2 sets up a vector holding the numbers $0, 1, 2, \ldots, 9$. Line 4 begins a *for loop*. The variable $i$ takes on the next value in the $t$ list each time through the loop. So, the first time in the loop, $i = t[0] = 0$, the second time, $i = t[1] = 1, \ldots$, and the 10th time, $i = 9$. Within the loop, the program simply prints the value of $i$. So the output looks like this.

**Output:**
```
[0 1 2 3 4 5 6 7 8 9]
0
1
2
3
4
5
6
7
8
9

Process finished with exit code 0
```

As with the *while loop*, we can include multiple statements to be executed in each pass through the loop by indenting the statements. Once the indentation ends, the loop definition is complete.

**Example.** Given $n = 5$, compute the following sum: $S = \frac{1}{1} + \frac{1}{2} + \frac{1}{3} + \cdots + \frac{1}{n} = \sum_{k=1}^{n} \frac{1}{k}$.

**Code:**
```
1   import numpy as np
2   n = 5
3   t = np.arange(1,n+1,1)
4   print(t)
5   S = 0
6   for k in t:
7       S = S+1.0/k
8   print('S =',S)
```

The $t$ values are set up to hold 1, 2, 3, 4, 5. The variable $S$ will hold the result of the sum and is initialized as zero. Then, in the first time through the loop, $k = 1$ and $S = S + 1/k = 0 + 1/1 = 1$.

The second time through the loop, $k = 2$ and $S = S + 1/k = 1 + 1/2 = 1.5$.

Likewise, the third time through the loop, $k = 3$ and $S = S + 1/k = 1.5 + 1/3 = 1.83333333$.

The fourth time through the loop, $k = 4$ and $S = S + 1/k = 1.83333333 + 1/4 = 2.08333333$.

And, finally, the fifth time through the loop, $k = 5$ and $S = S + 1/k = 2.08333333 + 1/5 = 2.283333333$.

Notice how $S$ accumulates the next term of the sum as it loops through the $k$ values. The program produces the following results.

**Output:**
```
[1 2 3 4 5]
S = 2.283333333333333
```

Try changing $n$ to 100. You should get the following final answer: $S = 5.187377517639621$.

Returning to our differential equation, we wish to loop through all of the $t$ values and compute the associated approximation values. When we calculated $\Delta t = \frac{b-a}{n}$, $n$ is the number of intervals of length $\Delta t$ in $[0, 5]$. There will actually be $n + 1$ total approximation points. Hence, we need an $y$ list with space to store $n + 1$ values as the approximations are computed. We wish to store all of the iterates so that we can plot them later or use them for computational purposes. One way to do this in Python is by creating a list of zeros as shown:

```
y = np.zeros(n+1).
```

Then, we will replace the zeros with the approximation values as they are computed. Thus, $y$ starts with $n + 1$ zeros. We are given the initial point. In this case, $(0, 1)$. That means that $y[0] = 1$. So we can assign that value immediately and use a for loop to compute the rest. Since we already know $y_0$, we need $y_1, y_2, \ldots, y_n$. Thus, we can set up an index list that has numbers 1 through $n$. Then, we use this list as the numbers to step through in the for loop. Here we go.

**Code:**

```
1    import numpy as np
2    import matplotlib.pyplot as plt
3
4    #solve dy/dt = (y-t)/2
5    #where when t=0, y=1
6
7    def rhs(t,y):
8        m = (y-t)/2
9        return m
10
11   #initial t value
12   a = 0
13   #final t value
14   b = 5
15   #number of intervals
16   n = 10
17   #delta t
18   dt = (b-a)/n
19
20   #create a vector of t-values
21   t = np.arange(a,b+dt,dt)
22   #create space for the y-values
23   y = np.zeros(n+1)
24   #create a list of indices
25   i = np.arange(1,n+1,1)
26   #we know the initial value of y to be 1
27   y[0] = 1
28   for k in i:
29       #compute the Euler approximation
30       #use the right hand side function to get the slope of the tangent line
31       m = rhs(t[k-1],y[k-1])
32       #get the next approximation
33       y[k] = m*dt+y[k-1]
34   #plot the solution
```

```
35   plt.plot(t,y)
36   plt.autoscale(enable=True, axis='x', tight=True)
37   plt.xlabel('t')
38   plt.ylabel('y')
39   plt.grid()
40   plt.show()
```

Line 25 creates a list of integers from 1 to $n$. These act as the index numbers for the loop, the $t$ values, and the $y$ values. The right-hand side of the DE is computed in line 31, and the recursive equation is imposed in line 33. The recursion stores the result in the appropriate location in the $y$ list. Finally, the program plots the approximation points to give a graph that represents the approximate solution of the differential equation. For $n = 10$, we obtain the following approximate solution.

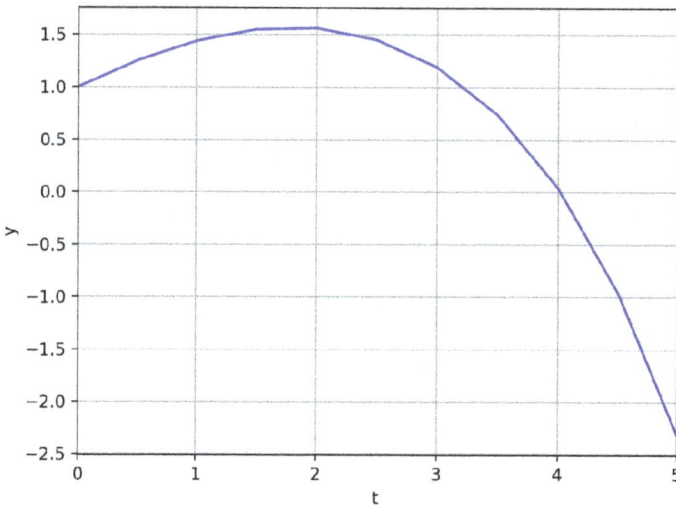

This particular differential equation can be solved analytically to find that the true solution is

$$y = t + 2 - e^{\frac{1}{2}t}.$$

The size of $\Delta t$ affects the accuracy of the approximation because $\Delta t$ indicates how far away from the point of tangency the approximation is. Larger values of $\Delta t$ then tend to cause more error because the approximation is further from the point of tangency. The smaller the value of $\Delta t$, the better the approximation. The next a graph shows the approximations for $n = 10$ and $n = 100$, along with the true solution. Note that the approximation with 100 points is significantly closer to the true solution.

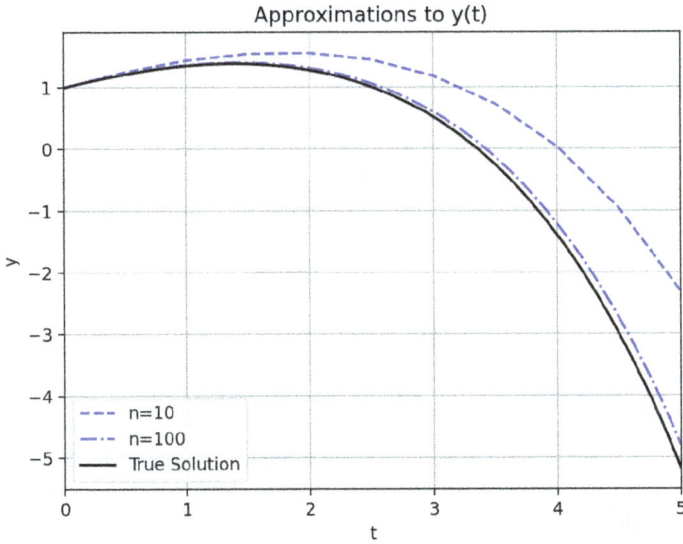

It must be said that Euler's method is a very simple method to solve particular types of differential equations, but many differential equations will be too complicated to use the method. Additionally, Euler's method is highly dependent on $\Delta t$ for accuracy. In fact, in many instances, if $\Delta t$ is too large, the method will fail. In such cases, there are many other available methods. Most of those methods are beyond the scope of this text, but the reader could investigate *implicit differential equation solvers* for more information or the *stability* of numerical methods for differential equations. Euler's method is straightforward and certainly demonstrates the concept of iteration.

Suppose we wish to modify Euler's method as follows. Instead of

$$y_{n+1} = y'(t_n, y_n)\Delta t + y_n,$$

we use

$$y_{n+1} = y'(t_{n+1}, y_{n+1})\Delta t + y_n.$$

In the first equation, $y_{n+1}$ is dependent only on the values that are already known, while in the second equation, $y_{n+1}$ depends on knowing the derivative at the next time step. Thus, we are trying to find the derivative at the next time step and the value of the function at the next time step simultaneously. This type of method is called an *implicit* method. We work our previous example using the modified Euler's method,

$$\frac{dy}{dt} = \frac{y-t}{2}, \quad y(0) = 1$$

$$y_{n+1} = y'(t_{n+1}, y_{n+1})\Delta t + y_n$$

$$y_{n+1} = \frac{y_{n+1} - t_{n+1}}{2} \Delta t + y_n.$$

Solving for $y_{n+1}$ gives

$$y_{n+1} = \frac{y_{n+1} - t_{n+1}}{2} \Delta t + y_n$$

$$2y_{n+1} = (y_{n+1} - t_{n+1})\Delta t + 2y_n$$

$$2y_{n+1} - y_{n+1}\Delta t = -t_{n+1}\Delta t + 2y_n$$

$$(2 - \Delta t)y_{n+1} = -t_{n+1}\Delta t + 2y_n$$

$$y_{n+1} = \frac{-t_{n+1}\Delta t + 2y_n}{2 - \Delta t}.$$

If we implement this scheme with Python, we can compare the results of the traditional Euler method with those just prescribed (we will call the new scheme the *implicit Euler* method). For $n = 20$, the comparison displayed in the next graph. For $n = 20$, the comparison is displayed in the next graph.

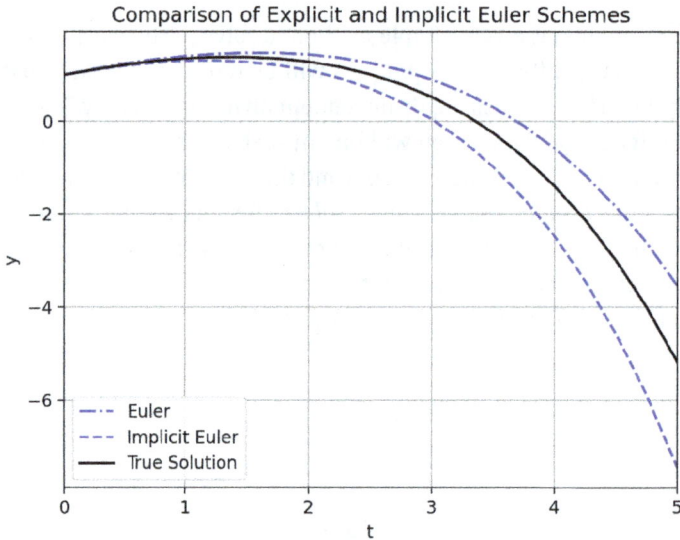

Comparison of Explicit and Implicit Euler Schemes

We can see that the implicit method differs from the explicit method. Implicit methods are usually more difficult to implement, sometimes requiring the use of matrices to solve a system of equations. However, implicit methods are generally more stable from a numerical perspective, having less dependence on $\Delta t$. The study of such stability could be encountered in a course in numerical analysis or computational analysis.

**See Exercises 6–7.**

### 5.2.1 Systems of differential equations and higher-order differential equations

In many cases, we are not trying to solve a particular differential equation, but, rather, a set or *system* of differential equations. Another issue that we have not addressed in any way is the presence of a second derivative (or higher) in differential equations. Surprisingly, if we can handle a system of equations, then we can handle many higher-order equations, as well.

Suppose, instead of

$$\frac{dy}{dt} = \frac{y - t}{2},$$

we have

$$\frac{dx}{dt} = y^2 - x$$
$$\frac{dy}{dt} = \frac{y - x}{2}$$

This is a *system of differential equations*. This particular system has two equations and two unknown functions, $y(t)$ and $x(t)$. We know that $x$ and $y$ are both functions because the derivatives indicate that the independent variable is $t$ (designated in the bottom of $\frac{dy}{dt}$ and $\frac{dx}{dt}$). The goal is to find approximations to these functions. Note that both $x$ and $y$ appear in both equations. Hence, we must somehow solve the equations simultaneously. As in our prior work, we will need some initial conditions in order to proceed such as an initial condition for each function. Let us assume that $y(0) = 1$ and $x(0) = 1.5$. Recall that in Chapter 4 we dealt with linear systems of equations by converting them into a matrix representation and then manipulating the matrix to achieve a desired form. Similarly, we will convert the system of differential equations into a matrix form. Then, we will generalize Euler's method so that it can be applied to the matrix representation. We begin by forming a column vector that contains our desired functions

$$\mathbf{Y} = \left[ \begin{array}{c} x(t) \\ y(t) \end{array} \right].$$

Then, we define the derivative of the vector to be the vector of the derivative of each component

$$\frac{d\mathbf{Y}}{dt} = \left[ \begin{array}{c} \frac{dx}{dt} \\ \frac{dy}{dt} \end{array} \right] = \left[ \begin{array}{c} x'(t) \\ y'(t) \end{array} \right].$$

Finally, we denote the right-hand side of the equations by

$$\mathbf{B} = \left[ \begin{array}{c} y^2 - x \\ \frac{y-x}{2} \end{array} \right].$$

So, the system can be presented in vector form as

$$\frac{d\mathbf{Y}}{dt} = \mathbf{B}(x, y, t).$$

This looks very similar to the single-equation form expressed earlier. The details from this point on would show that we need to apply Euler's method to each component of the vector form of the equation. As it turns out, the form of the iteration generalizes to vector form quite naturally. We find that

$$\mathbf{Y}_{n+1} = \mathbf{B}(x_n, y_n, t_n)\Delta t + \mathbf{Y}_n.$$

The code to solve the example system for $0 \le t \le 20$ is given below.

**Code:**

```
1   import numpy as np
2   import matplotlib.pyplot as plt
3
4   def rhs(t,yvec):
5       dy = np.zeros(2)
6       dy[0] = yvec[1]**2-yvec[0]
7       dy[1] = (yvec[1]-yvec[0])/2
8       return dy
9
10  #initial t value
11  a = 0
12  #final t value
13  b = 20
14  #number of intervals
15  n = 500
16  #delta t
17  dt = (b-a)/n
18
19  #create a vector of t-values
20  t = np.arange(a,b+dt,dt)
21  #create space for the y-values
22  y = np.zeros((n+1,2))
23  #create a list of indices
24  i = np.arange(1,n+1,1)
25  y[0,0] = 1.5
26  y[0,1] = 1
27  for k in i:
28      #compute the Euler approximation
29      #use the right hand side function to get the slope of the tangent line
30      dy = rhs(t[k-1],y[k-1,:])
31      #get the next approximation
32      y[k,:] = dy*dt+y[k-1,:]
33  #plot the approximations
```

```
34   plt.plot(t,y[:,0])
35   plt.plot(t,y[:,1])
36   #plot true solution
37   plt.autoscale(enable=True, axis='x', tight=True)
38   plt.xlabel('t')
39   plt.grid()
40   plt.legend(['x(t)','y(t)'])
41   plt.show()
```

We should take some time to discuss the differences between this code and the code we wrote for the single equation.

- First, in lines 4–8, we define the right-hand side function. In the single equation code, the arguments were values for the independent variable, $t$, and the dependent variable (desired function), $y$. In the system version, the second argument will contain a vector of values for each of the functions included in the system. The variable yvec will contain a value for $x(t)$ in the first component and a value for $y(t)$ in the second component.

- Similarly, while the single equation version returned a slope, $m$, the system version returns a vector, $dy$, which contains the slope of each of the functions.

- The variables $a, b, n$, and $\Delta t$ remain the same in the system version as in the single-equation version.

- In line 22, the matrix is initialized to store the approximation values for $x(t)$ and $y(t)$. Note that the matrix has as many rows as there will be approximations and a column for each function in the system (in this case, two). So, the first column will store the approximations for $x(t)$, and the second column will hold the approximations for $y(t)$.

- The $0^{th}$ row of $y$ corresponds to the initial conditions of the system. Thus, since $x(0) = 1.5$, the $0^{th}$ column is 1.5, $y[0,0] = 1.5$. Likewise, since $y(0) = 1$, the $1^{st}$ column of $y$ should be 1, $y[0,1] = 1$.

- Line 30 calls the rhs function to compute for each iteration. Notice the second argument is $y[k,:]$. This is the $k^{th}$ row of $y$, which is a vector containing two values (expected by rhs).

- Line 32 computes the values of the approximations for $x(t)$ and $y(t)$ for the $k^{th}$ time step (iteration) and assigns them to the $k^{th}$ row of $y$. Notice how similar this is to the single-equation version, $y[k] = m * dt + y[k-1]$. The slope $m$ is replaced by the vector of slopes, $dy$, and $y[k]$ is replaced by the vector at the $(k-1)^{st}$ level. Otherwise, the iteration is the same.

- Once the for loop is complete, the approximations for both functions are stored in $y$. Line 34 plots the first column of $y$, which represents $x(t)$, while line 35 plots the second column, which represents $y(t)$.

This generates the following graph.

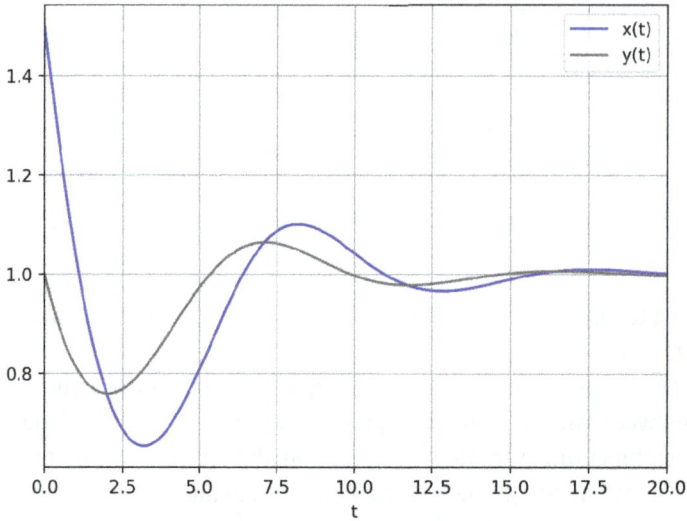

Notice that both $x(t)$ and $y(t)$ seem to approach 1 as $t$ gets large. This is an example of an equilibrium point. We say that the system would reach a *steady state* at the point ($x = 1$, $y = 1$).

In systems containing two or three equations (and, thus, two or three functions), it is common to examine how the functions behave simultaneously. We do this by plotting what is known as a *phase portrait*. You can think of a phase portrait as the plot of a path where the points of the path are given by the function values. In our example, we could think of $x(t)$ as the $x$-coordinate and $y(t)$ as y-coordinate. The phase portrait would look like this.

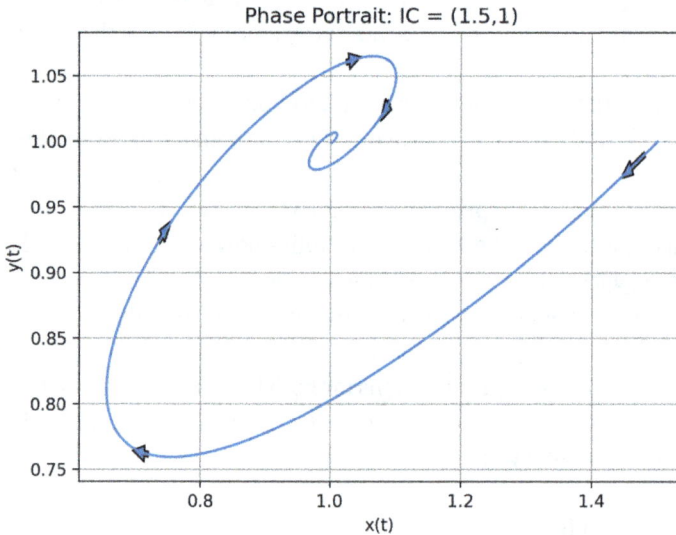

The arrows placed on the curve show the direction in which travel as time ($t$) increases. We can tell that the curve starts at $(1.5, 1)$, which corresponds to the initial conditions of $x$ and $y$, and spirals to the point $(1, 1)$, which corresponds to the steady state. Also, the arrows are uniformly spaced with respect to $t$. So, we can see that the speed along the path is faster (more distance traveled in the same amount of time) in the beginning than it is when it approaches the steady state. The speed would be zero once the system reaches steady state. The phase portrait allows us to know how the two curves interact (and it is a really cool picture). Except for the directional arrows, the phase portrait is accomplished by plotting the values for the $x$ function and the corresponding values of the $y$ function. The arrows are helpful, but not a focal point in this text. For completeness, the code to plot the phase portrait, including the arrows, is given next.

**Code:**

```
1   plt.figure()
2   plt.plot(y[:,0],y[:,1])
3   head = 1
4   tail = 0
5   w = 55
6   dx = y[head,0]-y[tail,0]
7   dy = y[head,1]-y[tail,1]
8   plt.arrow(y[head,0],y[head,1],dx,dy,width=.004)
9   numarrows = int((n-head)/w)
10  for i in range(4):
11      head = head + w
12      tail = tail + w
13      dx = y[head,0]-y[tail,0]
14      dy = y[head,1]-y[tail,1]
15      plt.arrow(y[head,0],y[head,1],dx,dy,width=.004)
16  plt.xlabel('x(t)')
17  plt.ylabel('y(t)')
18  plt.title('Phase Portrait: IC = (1.5,1)')
19  plt.grid()
20  plt.show()
```

There are many technical details that we are not covered here. In fact, for certain initial conditions, the method would fail, and, it is possible the system may not be solvable. If such topics are of interest, the reader may wish to pursue courses in *numerical analysis* and *differential equations*.

To close our discussion of systems of differential equations, we turn our attention to higher-order differential equations. Consider the following equation:

$$\frac{d^2y}{dt^2} - \mu(1 - y^2)\frac{dy}{dt} + y = 0.$$

This equation is called the *van der Pol equation* and arises in the study of circuits containing vacuum tubes. We call the equation a second-order equation because the highest order derivative is the second derivative. So, what does this have to do with systems of differential equations? Our strategy will be to transform the higher-order equation into a system of first-order equations. Then, we can use the methods previously discussed. To make the transformation, we introduce a second function, say $x(t)$, and define it to be the derivative of $y(t)$. That is, $x(t) = y'(t)$. Then, $x'(t) = y''(t)$. Thus, the van der Pol equation could be written as

$$x'(t) - \mu(1 - y^2)x(t) + y(t) = 0$$
$$y'(t) = x(t).$$

Solving the first equation for $x'$ gives

$$x'(t) = \mu(1 - y^2)x - y$$
$$y'(t) = x(t).$$

Thus, the original second-order equation is expressed as a system of two first-order equations. Then, we can use Euler's method for systems to approximate the solution to $x(t)$ and $y(t)$. Note that the approximations that are of most interest are those for $y(t)$ since that is the only function in the original equation. As in previous examples, we will need initial conditions for each of the functions. Generally, these conditions are dictated via the context of the problem. In this abbreviated academic setting, we do not have such context. For purpose of illustration, assume that $x(0) = 1$, $y(0) = 1$, and $\mu = 1$. Solving the system for $0 \leq t \leq 20$ gives the following graphs for $x(t)$ and $y(t)$.

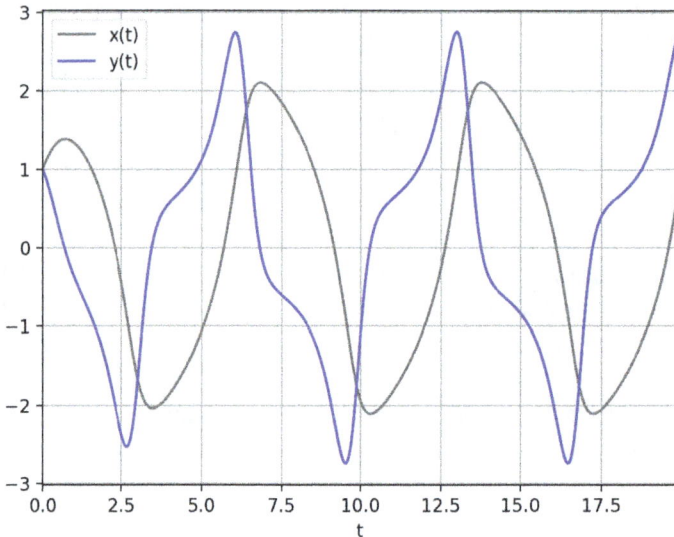

The function $y(t)$ is called a *van der Pol oscillator* that we can see is a periodic function. For any smooth, periodic function, the derivative would also have to be periodic with the same period (why?). It is clear this is the case from the graph of $x(t)$. Since the functions are periodic, the phase portrait would also become periodic. This means that it would trace out the same loop repeatedly once the functions have established their periodicity. The phase portrait of the system is depicted in the next graph.

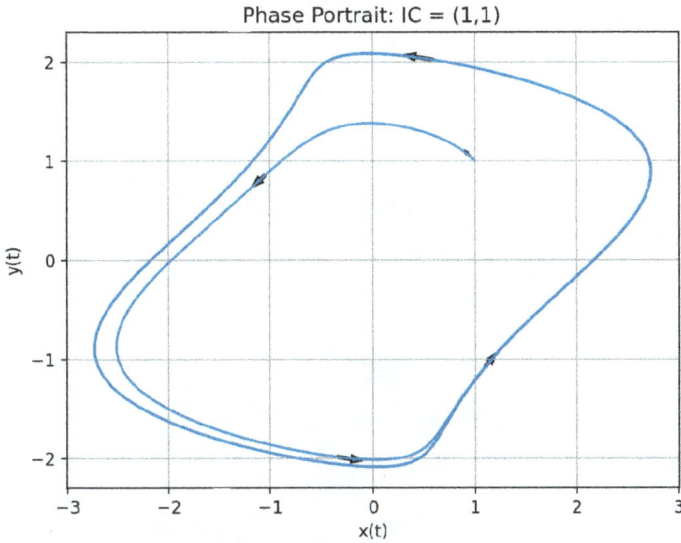

Phase Portrait: IC = (1,1)

Here is the code to compute the approximations and produce the graphs.

**Code:**

```
1   import numpy as np
2   import matplotlib.pyplot as plt
3
4   def rhs(t,yvec):
5       mu = 1
6       dy = np.zeros(2)
7       dy[0] = mu*(1-yvec[1]**2)*yvec[0]-yvec[1]
8       dy[1] = yvec[0]
9       return dy
10
11  #initial t value
12  a = 0
13  #final t value
14  b = 20
15  #number of intervals
```

```
16   n = 500
17   #delta t
18   dt = (b-a)/n
19
20   #create a vector of t-values
21   t = np.arange(a,b+dt,dt)
22   #create space for the y-values
23   y = np.zeros((n+1,2))
24   #create a list of indices
25   i = np.arange(1,n+1,1)
26   y[0,0] = 1
27   y[0,1] = 1
28   for k in i:
29       #compute the Euler approximation
30       #use the right hand side function to get the slope of the tangent line
31       dy = rhs(t[k-1],y[k-1,:])
32       #get the next approximation
33       y[k,:] = dy*dt+y[k-1,:]
34   #plot the approximations
35   plt.plot(t,y[:,0])
36   plt.plot(t,y[:,1])
37   #plot true solution
38   plt.autoscale(enable=True, axis='x', tight=True)
39   plt.xlabel('t')
40   plt.grid()
41   plt.legend(['x(t)','y(t)'])
42   plt.figure()
43   plt.plot(y[:,0],y[:,1])
44   head = 1
45   tail = 0
46   w = int(n/12)
47   dx = y[head,0]-y[tail,0]
48   dy = y[head,1]-y[tail,1]
49   plt.arrow(y[head,0],y[head,1],dx,dy,width=.01)
50   numarrows = int((n-head)/w)
51   for i in range(4):
52       head = head + w
53       tail = tail + w
54       dx = y[head,0]-y[tail,0]
55       dy = y[head,1]-y[tail,1]
56       plt.arrow(y[head,0],y[head,1],dx,dy,width=.025)
57   plt.xlabel('x(t)')
58   plt.ylabel('y(t)')
59   plt.title('Phase Portrait: IC = (1,1)')
```

```
60    plt.grid()
61    plt.show()
```

**See Exercise 8.**

### 5.2.2 Interpolation—using the approximations

In the previous section, we obtained approximate values of a van der Pol oscillator at particular values of $t$. In our example, we solved for $0 \leq t \leq 20$ with $n = 500$. So we have 501 approximations of $y$ for $t$ values that begin at 0 and are separated by 0.04 units. The first ten approximation points are given in this table.

| $t$ | 0.000 | 0.040 | 0.080 | 0.120 | 0.160 | 0.200 | 0.240 | 0.280 | 0.320 | 0.360 |
|---|---|---|---|---|---|---|---|---|---|---|
| $y$ | 1.000 | 1.040 | 1.078 | 1.115 | 1.150 | 1.182 | 1.212 | 1.240 | 1.266 | 1.289 |

One question that arises is: 'What if we need $y(c)$ where $c$ is not one of the $t$ values for which we have an approximation?' Suppose we wished to approximate $y(0.0732)$? The most common way to find such an approximation is by using *linear interpolation*. The idea is to construct a linear function that connects the known approximation points that surround the $t$ value of interest. Then use the linear function to find the desired value of $y$. For $t = 0.0732$, the surrounding $t$ values would 0.040 and 0.080. Thus, the corresponding $(t, y)$ pairs are $(0.040, 1.040)$ and $(0.080, 1.078)$. We use algebra to find the equation of the line containing these points.

$$m = \frac{y_2 - y_1}{t_2 - t_1}$$
$$m = \frac{1.078 - 1.040}{0.08 - 0.04}$$
$$m = 0.95$$

point–slope form
$$y - y_1 = m(t - t_1)$$
$$y - 1.040 = 0.95(t - 0.04)$$
$$y = 0.95t + 1.002$$

Now, we can approximate $y(0.0732)$ by substituting $t = 0.0732$ into the linear function that we just found. So,

$$y(0.0732) \approx 0.95(0.0732) + 1.002 = 1.07154.$$

We can use this interpolation when we need to find approximations within the range of those that have already been computed. The following code presents a crude function to perform this interpolation.

**Code:**

```
1   import numpy as np
2
3   # x is the new input value, t is the vector of x-values
4   # y is the vector of y-values
5   def interp(x,t,y):
6       n = len(t)
7       startindex = 0
8       # find the indices between which the new x value lies
9       while t[startindex]<x:
10          startindex = startindex + 1
11      startindex = startindex - 1
12      endindex = startindex +1
13      # slope for interpolation
14      m = (y[endindex]-y[startindex])/(t[endindex]-t[startindex])
15      # compute approximation using point slope form
16      y_of_x = m*(x-t[startindex])+y[startindex]
17      return y_of_x
18
19  # begin main program
20  t = np.array([0.000, 0.040, 0.080, 0.120, 0.160,\
21              0.200, 0.240, 0.280, 0.320, 0.360])
22  y = np.array([1.000, 1.040, 1.078, 1.115, 1.150,\
23              1.182, 1.212, 1.240, 1.266, 1.289])
24
25  #approximate y(0.0732)
26  x = 0.0732
27  yinterp = interp(x,t,y)
28  print('y({}) = {:.5f}'.format(x,yinterp))
```

**Output:**
```
y(0.0732) = 1.07154

Process finished with exit code 0
```

**See Exercise 9.**

## 5.3 Exercises

1.  (a) Modify the `bisect` function so that, in addition to the approximate solution, the number of iterations needed to achieve the indicated tolerance is also returned.
    (b) Plot the function $f(x) = x^3 - 100\cos(x)$ on the interval $[-1, 4]$.
    (c) Find the root of the function using a tolerance of 0.0001.
2.  (a) Change the loop condition in `bisect` so that the iteration stops when the difference between two successive iterates is less than a designated tolerance.
    (b) Find the same root as that found in problem 1. Compare the number of iterations for each type of stopping criteria.
    (c) Which stopping method do you think is better?
3.  The population of a certain bacteria follows the form of the following function:

$$P(t) = \begin{cases} e^t & \text{if } 0 \leq t < 3 \\ \frac{at+3}{t+5} & \text{if } t \geq 3. \end{cases}$$

Use the bisection algorithm to find the value of $a$ that makes the function continuous.

4.  Find the maximum value of $f(x) = e^x \sin(x) - \frac{x^2}{2} + 5$ when $-1 \leq x \leq 3$.
5.  Another method to find roots that we learned in the first semester of calculus is *Newton's method*. Write a Python function to implement Newton's method for finding roots. Then, repeat problem 1 using Newton's method with an initial guess of $x_0 = 1$. Compare the number of iterations for each of the two methods. What happens if the initial guess is changed to $x_0 = -1$?
6.  (a) Modify the code for the explicit Euler scheme so that it is defined as a function. What arguments should be included?
    (b) Use the function to solve the differential equation

$$\frac{dy}{dx} = x^2 - \sin(x).$$

Use a step size of .1 on the interval from 0 to $2\pi$ with $y(0) = 0$. Graph the approximate solution.

    (c) We can solve the previous equation analytically to find that $y = \frac{x^3}{3} + \cos(x) - 1$. Plot the true solution on the same graph as the approximation.

7.  Write a Python program to perform both the explicit and implicit Euler methods to solve

$$\frac{dy}{dt} = \cos(t) + e^{-t}y$$

for $0 \leq t \leq 20$. Let $y(0) = 0$ with a step size of 0.05 on the interval from 0 to $2\pi$. Plot both approximations on the same axes.

8. A relationship that commonly occurs in nature is that of a predator and its prey. Under certain assumptions, the populations of a predator and prey can be modeled by the system of differential equations

$$\frac{dX}{dt} = aX - bXY$$

$$\frac{dY}{dt} = cY + dXY,$$

where $X$ represents the population of the prey, $Y$ represents the population of the predator, and $a$ and $b$ represent the birth rate and death rate and $c$ and $d$ interaction parameters.

Solve the system of differential equations using Euler's method on the time interval from $t = 0$ to $t = 365$ ($t$ measured in days). Use trial and error to determine an appropriate value for $n$ (and hence, $\Delta t$). Use the following values:

$$a = 0.04$$
$$b = 0.0005$$
$$c = -0.1$$
$$d = 0.0005$$
$$X(0) = 50$$
$$Y(0) = 10.$$

   (a) On the same graph, plot the populations of predator and prey throughout the time period. Are the populations periodic? Use 50,001 points ($\Delta t = \frac{t_{final} - t_0}{50000}$). Why or why not is that the case?
   (b) Use the approximations of $X$ and $Y$ to draw a phase portrait. Using the phase portrait, explain the cycle of the populations of each species. You do not need to put arrows in the phase portrait.

9. Suppose we have the ten approximation points given at the beginning Section 5.2.2. Let **t** be a vector of time values consisting of 76 evenly spaced values between 0 and 0.36, i. e.,

$$\mathbf{t} = [0, .0048, .0096, \ldots, 0.36].$$

   Write a Python program to find the approximation associated with each $t$ value and plot the original set of ten points and the new set of 76 points on the same graph.

# 6 Statistics

We turn our attention now to projects that analyze large data sets. There are two general types of project that we will consider:
- Projects that use data to infer conclusions about a larger group, called a population.
  - This is usually done with hypothesis testing or confidence intervals.
- Projects that use data to predict results based on known results within the data.
  - This is usually accomplished with some type of regression.

We will study both of these scenarios later in the chapter, but, first, we must discuss how to obtain the data for such projects.

## 6.1 File handling

When dealing with large amounts of data, we must have ways of retrieving, filtering, and saving the data. This is typically done via data files. Thus, we need to understand how Python can open and close data files and how it can read data from a file and write data to a file. One of the most common types of data files is the *comma separated value* file. It uses *.csv* as the extension to the filename. The file is formatted so that data values are separated by commas. Each line of the file corresponds to a data record that includes all the values associated with one instance or observation of an experiment. As an example, we found data that provides hourly surface-climate measurements from 122 weather stations in Brazil. The data file can be found here: https://www.kaggle.com/PROPPG-PPG/hourly-weather-surface-brazil-southeast-region. The filename is *brazilclimate.csv*. One of the first things we might try when we find a data file of interest is to open it with Microsoft Excel. Excel can open .csv files, but, in this case, the file is too large for Excel to load all of the data it contains. This gives us our first reason to use something like Python to analyze the data. However, loading just a portion of the file with Excel does allow us to see the types of data contained in the file. A screen shot of the Excel window is shown below in Figure 6.1. Examining the spreadsheet allows one to discover several things about the data:
- there are 31 columns, indicating 31 different measurements for each row;
- the sheet contains the column headings
  - we can make sense of many of the headings, but some we may need to investigate (go back to the data source);
- there are a variety of types of data, including numbers, dates, and strings;
- it is clear that there is some missing or errant data in the sheet
  - there are some zeros that seem to be incorrect, and there are many cells left blank.

https://doi.org/10.1515/9783110776645-006

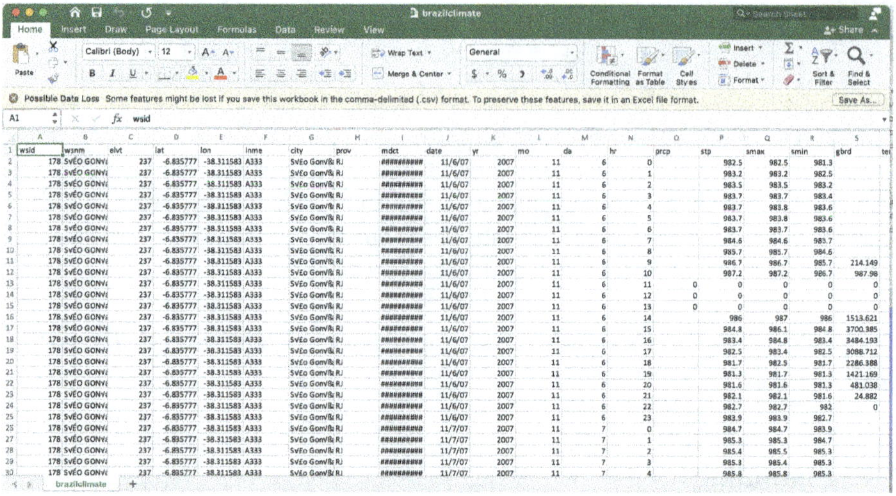

**Figure 6.1:** Excel spreadsheet showing a portion of the data file.

We need to address the issues with the data as best we can. Sometimes, we will discard records, and sometimes we can correct them. The process of "fixing" issues with the data is called *data cleaning*. For example, the data has a column called *prcp*. In the spreadsheet, it seems as if the value in this column is either absent or zero. Is this always the case? If so, we can discard the column. So let's investigate to see if the *prcp* is ever nonzero. To do so, we need to open the file and read through every value of *prcp* to see if it is ever nonzero. For simplicity, we will store the file in the same folder as the Python program. If the data file is in a different folder, then we will have to use a file path to indicate where the file is located. Consider the following code.

**Code:**

```
1  climate_file = open('brazilclimate.csv','r')
2  count = len(climate_file.readlines())
3  print('Number of lines in the file is',count)
4  climate_file.close()
```

Line 1 of the code executes an open statement. This statement opens the file for processing. The 'r' in the second argument indicates that the file is allowed to be read, but not written to. Files can be opened for reading (r), writing (w), both reading and writing (rw), or for appending (a). We will discuss writing files shortly. The file is assigned a file identifier (fid) of `climate_file`. The fid name can be any valid variable name. In line 2, `climate_file.readlines()` reads every line of the file and stores the information in the computer's memory. The `len` function returns the number of lines that were read. Line 3 then prints out how many lines are contained in the file. Finally, any file that is opened should be closed before the program ends. Hence, line 4 closes the file. The output of the code follows.

**Output:**
```
Number of lines in the file is 9779169

Process finished with exit code 0
```

We see that there are nearly 10 million records of data in the file. For the purposes of illustration, a second file has been created called *smallclimate.csv* which contains only the first 3,000 lines of *brazilclimate.csv*. We will work with the smaller file because it allows us to show more meaningful output. For example, if we decide to show all of the lines for which *prcp* is nonzero, the larger file may contain thousands of such records. Displaying all of those would not be helpful. To see if *prcp* contains any nonzero values, we need to look at each record to see what the value of *prcp* is. This requires us to break each line of data into its separate fields (columns). Rather than read all the lines at the same time (.readlines()), we will read one line at a time, find the *prcp* column, get its value, and decide if the value is nonzero. One way to accomplish this is with the following code.

**Code:**
```
1   climate_file = open('smallclimate.csv','r')
2   #we know the first line of this file contains headers of the columns
3   record = climate_file.readline()
4   #set up a counter to know what line we are on
5   count = 1
6   #set up a counter to count number of nonzero values for prcp
7   num_non0 = 0
8   #read lines until you reach a blank line, then assume you are done
9   while record != '':
10      record = climate_file.readline()
11      #split the record into its separate columns
12      record_vector = record.split(',')
13      #prcp is the 15th column.  in Python, that is index 14
14      #convert from a string to value
15      if record_vector[14] == '':
16          prcp = 0
17      else:
18          prcp = float(record_vector[14])
19          if prcp != 0.0:
20              num_non0 = num_non0 + 1
21              print('prcp = {:.4f} in record #{}.'.format(prcp,count))
22      count = count + 1
23  print('The number of nonzero values is {}.'.format(num_non0))
24  climate_file.close()
```

The output of the program follows.

**Output:**
```
Traceback (most recent call last):
  File "/Users/WillMiles/Desktop/_Courses/SciComp/SciCompBook/BookCode/
        basics4.py", line 18,
  in <module>
    if record_vector[14] == '':
IndexError: list index out of range
prcp = 0.6000 in record #1726.
prcp = 0.2000 in record #1860.
prcp = 0.2000 in record #1980.
.

.

.
prcp = 0.2000 in record #2980.


Process finished with exit code 1
```

This small program has much to show us about accessing information in files, as well as an important error that we need to be aware of. We thoroughly explain this program step-by-step.

*Line 1:* Opens the data file for reading.

*Line 3:* Since we know the first line of the file contains the headers, we read it so that our file pointer is actually pointing to the second line. There is no need to find the *prcp* value in the header line. Notice that the method is `.readline()`, not `.readlines()`. The line from the file is stored as one long string in the variable `record`.

*Line 5:* We may want to know which line of the file is currently begin accessed. Thus, we set up a counter in the program called `count` to keep track of the location in the file.

*Line 7:* Likewise, we will want to know how many nonzero values are found for *prcp*. Thus, a counter is initialized at 0 and will be updated as we find nonzero values. The number of nonzero values is to be stored in the variable `num_non0`.

*Line 9:* We use a while loop to read each line of the file. The presumption is that, if a line is blank, then we must be at the end of the file, and the code should exit the loop. We test for the condition with `record != ''`. However, we will see that this assumption is incorrect.

*Line 10:* This line reads the next record (line) in the file. The record is stored as one long string that includes the commas separating the data fields.

*Line 12:* In order to split the long string retrieved in line 10 into the separate fields (or columns), we use the `.split` method. The command `record.split(',')` searches through the string in `record` for commas. It uses commas to delineate separate fields in the string and creates a vector with each field as an element of the vector. For example, the line from the file may look like this:

```
178,SÃO GONÇALO,237.00,-6.835777,-38.311583,A333,São Gonçalo, ...
```

Once the split has been executed, `record_vector` would be:

```
['178', 'SÃO GONÇALO', '237.00', '-6.835777', '-38.311583', 'A333',
 'São Gonçalo', ...]
```

Notice that the elements of `record_vector` are strings.

*Lines 15–18:* These few lines make the decisions on whether the *prcp* field has a nonzero value. Since the *prcp* field is in the $15^{th}$ column, it would be in the $14^{th}$ index of `record_vector`. So, testing if `record_vector == ''` is checking to see if the value is blank. If it is, then we can likely assume the value is zero. Thus, we assign a variable named `prcp` to be zero. If the field is not blank, then we drop to the else portion of the if block. Since we know the field is not blank, we need to determine the value in the field. The current contents of the field are in string form. Hence, we need to convert the string to a floating-point number. This is done by using the float function applied to `record_vector[14]` and assigning the resulting value to `prcp` in line 18. At this point, `prcp` holds the value of column 15. If the column is blank, the value is zero.

*Lines 19–21:* In line 19, we test to see if the value in `prcp` is nonzero. If it is, we increment the count of nonzero entries (line 20) and print the amount on the screen (line 21). Since precipitation cannot be negative, it may be worth modifying the code slightly to see if the value of `prcp` is negative or positive. A negative value would indicate an error in the data.

*Lines 22:* At the bottom of the while loop, we increment `count` to indicate that we are moving to the next line of the file.

*Lines 23:* Once the while loop has been completed, the code displays the total number of nonzero entries that were found.

*Lines 24:* The data file is closed, and the program ends.

Now, from the output, we see that something has gone wrong. There is a message that indicates there is a problem with the statement: `if record_vector[14] == '':`. More precisely, the message indicates that a 'subscript is out of range.' This means that there were not 15 entries in this record, but the loop condition terminates the loop if the record is blank. The issue is that the record is read at the beginning of the while loop. So, the last record is read, and the code tries to process the record before the while statement tests to see if the record is blank. We need to move the .readline command to the bottom of the while loop. However, this would then cause issues with the first record. We have read the header line and would have errors trying to convert the header *prcp* into a number. Thus, we need to read the second record before the while loop begins. Then, the code should run completely. The revised code and the output are given here.

**Code:**

```
1   climate_file = open('smallclimate.csv','r')
2   #we know the first line of this file contains headers of the columns
3   record = climate_file.readline()
4   #set up a counter to know what line we are on
5   count = 1
6   #set up a counter to count number of nonzero values
7   num_non0 = 0
8   # read the second line (which is the first line containing actual data)
9   record = climate_file.readline()
10  #read lines until you reach a blank line, then assume you are done
11  while record != "":
12      record = climate_file.readline() #move to bottom of the loop
13      #split the record into its separate columns
14      record_vector = record.split(',')
15      #prcp is the 15th column.  in Python, that is index 14
16      #convert from a string to value
17      if record_vector[14] == '':
18          prcp = 0
19      else:
20          prcp = float(record_vector[14])
21          if prcp != 0.0:
22              num_non0 = num_non0 + 1
23              print('prcp = :.4f in record #.'.format(prcp,count))
24      count = count + 1
25      record = climate_file.readline()
26  print('The number of nonzero values is .'.format(num_non0))
27  print('There were  records processed.'.format(count))
28  climate_file.close()
```

The changes to the code are in blue.

**Output:**

```
prcp = 0.6000 in record #1726.
prcp = 0.2000 in record #1860.
```

```
prcp = 0.2000 in record #1980.
.

.

.
prcp = 0.2000 in record #2980.
The number of nonzero values is 39.
There were 3000 records processed.

Process finished with exit code 0
```

We can now see that the code runs without error. There were 39 nonzero values in the 3,000 records. If *prcp* is actually a quantity that we wish to analyze, we would likely want to write a new data file that had the zeros written in place of the blanks. We will address this shortly. Suppose we have a project that proposes to study the precipitation and maximum temperature based on the time of year and location. Then, we need very little of the data that is included in the *brazilclimate.csv* file. So, we will create a new file that is "clean" and contains only the data of interest.

We still want to replace the blanks with zeros in the prcp column, and, in addition, we want to examine the temperature, date, and station ID fields as well. When we examine the headers, we can determine which columns are needed:

| Field Name | Column |
| --- | --- |
| Station ID | 1 |
| Latitude | 4 |
| Longitude | 5 |
| Year | 11 |
| Month | 12 |
| Day | 13 |
| Hour | 14 |
| Precipitation | 15 |
| High Temperature | 22 |

We will deal with blanks in precipitation as before. If the station ID or temperature is blank, then we have other issues. Since there is no definitive way to deal with this possibility, we will discard any records that have such issues. We can fix latitude and longitude blanks if we have the station ID (we can just use the lat and long values for another record with the same station ID). We have to decide what to do if we have blanks in the date fields. For now, we will discard the record if the year or month is blank. Finally, we will store our cleaned data in a file called *tempstudy.csv*. We will process the larger file this time. So we will read data from *brazilclimate.csv* and write data to *tempstudy.csv*. Let's determine a plan that will process the file as desired using the following options.

- If station id, high temperature, year, or month is blank, discard the record.
- If latitude or longitude is blank and station id is blank, discard the record.
- If latitude or longitude is blank and station id is not blank, then find the correct latitude and longitude
    - We will test to see if any records meet this criterion before writing the logic if needed.
- If precipitation is blank, then use a value of zero.

The code to process the file follows.

**Code:**

```
1   # open the climate file for reading
2   climate_file = open('brazilclimate.csv','r')
3   # open the new (output) file for writing. If the file does not exist,
4   # it is created.
5   temperature_file = open('tempstudy.csv','w')
6   # write the headers to the output file.  the \n is a next line indicator
7   headerline = 'ID,lat,long,year,month,day,hour,precip,temp\n'
8   temperature_file.write(headerline)
9   # we know the first line of the input file contains headers of the columns
10  record = climate_file.readline()
11  # set up a counter to know what line of the input file we are on
12  count = 1
13  # set up counters to count the number of records that are corrected and
14  # the number of records that are discarded
15  corrected_recs = 0
16  discarded_recs = 0
17  # get the first non-header line (this is the first line with actual data)
18  record = climate_file.readline()
19  # read lines until you reach a blank line, then assume you are done
20  while record != "":
21      # initialize a variable (discard flag) to indicate whether
22      # the record should be discarded
23      # 0 = keep the record (write to the new file),
24      # -1 = discard the record (do not write to the new file)
25      # discarded records are not deleted from original file
26      # initialize the flag to "keep"
27      discardflag = 0
28      # this just lets me know that the program is progressing through the file
29      # by printing to the screen every millionth record
30      if count%1000000 == 0:
31          print(count)
32      # split the input record into its separate columns
33      record_vector = record.split(',')
```

```
34    # check station ID, year, month, temperature for blanks the \ allows us to
35    # continue the code line to the next line for readability
36    if record_vector[0] == '' or record_vector[10] == '' \
37            or record_vector[11] == '' or record_vector[21] == '':
38        # if any of them is blank, set the discard flag and increment the
39        # the discarded records counter
40        discardflag = -1
41        discarded_recs = discarded_recs + 1
42    # see if there are any blank lat/long values that have a station id
43    if record_vector[3] == '' or record_vector[4]=='':
44        if record_vector[0] != '':
45            print('need to look up station id')
46        else:
47            # if the lat and long and id are all blank, discard the record
48            discardflag = -1
49    # now replace blank precip values with 0.  only need to do this in the new
50    # file
51    if record_vector[14] == '':
52        record_vector[14] = '0'
53        # increment corrected records counter
54        corrected_recs = corrected_recs + 1
55    # build and write the record to the output file
56    if discardflag == 0:
57        # temp_record holds the record to be written
58        # the + will just concatenate the strings.  the \ is a line
59        # continuation character
60        # we are building a string with all the desired
61        # fields separated by commas
62        temp_record = record_vector[0]+','+record_vector[3]+','\
63        +record_vector[4]+','+record_vector[10]+','+record_vector[11]+','\
64        +record_vector[12]+','+record_vector[13]+','+record_vector[14]+','\
65        +record_vector[21]+'\n'
66        # write to the output file
67        temperature_file.write(temp_record)
68    # read the next line of input
69    record = climate_file.readline()
70    count = count+1
71    # go back to the top of the loop
72 # the loop is complete
73 # close the files
74 climate_file.close()
75 temperature_file.close()
76 # print the counts
77 print('Number of discarded records:',discarded_recs)
78 print('Number of corrected records:',corrected_recs)
```

The output is shown below.

**Output:**
```
1000000
2000000
3000000
4000000
5000000
6000000
7000000
8000000
9000000
Number of discarded records: 26
Number of corrected records: 8371184

Process finished with exit code 0
```

The code is commented extensively, and the reader is encouraged to step through the code, especially the if structures, to see how the code addresses the possible issues in the file. The fact that 'need to look up station id' is never displayed, indicates that we do not have to address the latitude and longitude issues because there are none. Note that the shorter record is written to the file *tempstudy.csv* in line 67. When the file is opened in line 5, it will be created if it does not exist. The file will be rewritten each time the code is executed. If we wished to add to an existing file, we would open it for append using the 'a' option in the open statement. The first few lines of *tempstudy.csv* are displayed here.

***tempstudy.csv***
```
ID,lat,long,year,month,day,hour,precip,temp
178,-6.835777,-38.311583,2007,11,6,0,0,29.7
178,-6.835777,-38.311583,2007,11,6,1,0,29.9
178,-6.835777,-38.311583,2007,11,6,2,0,29.0
178,-6.835777,-38.311583,2007,11,6,3,0,27.4
178,-6.835777,-38.311583,2007,11,6,4,0,26.3
.
.
.
```

Now that we now have the cleaned data ready for use, let's write a program to calculate the average precipitation and the average temperature for each month of the year. There are many ways to accomplish this. Our approach will be as follows.
1.   Read in each line of the tempstudy.csv file.

2. For each record, determine the month for the data, the precipitation amount, and the high temperature.
3. Add the precipitation amount to the current precipitation amount for that month.
4. Add the high temperature to the total temperature for that month.
5. Increment a counter to learn how many observations apply to that month.
6. When all records have been read, divide the totals by the number of observations for each month.

The program is given here.

**Code:**

```
1   import numpy as np
2   import matplotlib.pyplot as plt
3   np.set_printoptions(precision=3,suppress=1,floatmode='fixed')
4
5   # open the temp study file for reading
6   temperature_file = open('tempstudy.csv','r')
7   # we know the first line of this file contains headers of the columns
8   record = temperature_file.readline()
9   # set up a matrix to hold what we need
10  # we need a row for each month and columns for
11  # total precip, total temperature, and number of observations
12  tempsummary = np.zeros((12,3))
13  # read lines until you reach a blank line, then assume you are done
14  temprec = temperature_file.readline()
15  count = 0
16  while temprec != '':
17      if count%1000000 == 0:
18          print(count)
19      # split the record into fields
20      tempvec = temprec.split(',')
21      # get the month, precipitation, and temperature for this observation
22      mth = int(tempvec[4])
23      precip = float(tempvec[7])
24      temp = float(tempvec[8])
25      tempsummary[mth-1,0] = tempsummary[mth-1,0] + precip
26      tempsummary[mth-1,1] = tempsummary[mth-1,1] + temp
27      tempsummary[mth-1,2] = tempsummary[mth-1,2] + 1
28      temprec = temperature_file.readline()
29      count = count + 1
30  temperature_file.close()
31  tempsummary[:,0] = tempsummary[:,0]/tempsummary[:,2]
32  tempsummary[:,1] = tempsummary[:,1]/tempsummary[:,2]
33  print(tempsummary)
```

Lines 1–3 are the usual import and formatting statements that we have used before. Line 6 opens the file that we created with the previous program. This time the file is opened for reading. As in other examples, the header line is read so that the file pointer is actually looking at data records. To compute the averages, we need to sum all of the precipitation amounts and all of the temperature amounts, and, then, divide by the number of observations. We will use a matrix to hold the necessary information. Line 12 establishes the matrix. Since we want the average for each month, we will have 12 rows in the matrix. Row 0 corresponds to January, row 1 to February, and so on. We will use the $0^{th}$ column to accumulate the precipitation, the $1^{st}$ column to accumulate the temperature, and the $2^{nd}$ column to keep track of the number of observations. Thus, the matrix is $12 \times 3$ and is initialized to all zeros. Line 14 reads the first actual data record. We enter the while loop with the data record. Line 20 splits the data record into its component fields (all fields are strings at this point). Lines 22–24 convert the month, precipitation, and temperature values in the record into numbers. The row of the matrix that is to be updated is one less than the month. Line 25 takes the current value in column 0 and adds the new precipitation value to it. Thus, it is summing all of the precipitation values. Line 26 does the same thing for temperature. Line 27 sums the number of records (observations) for that month. Once the while loop is complete, the tempsummary matrix has all of the components needed to compute the averages. The file is closed because we will not need it anymore.

To compute the averages, we could establish another matrix, or we can reuse the one we have. tempsummary currently has the following information in each row:

| Column 0 | Column 1 | Column 2 |
| --- | --- | --- |
| sum of all precipitation values for the month | sum of all the temperature values for the month | total number of observations for the month |

Thus, we need to take columns 0 and 1 and divide them by column 2. This is done in lines 31 and 32. In line 31,

```
tempsummary[:,0] = tempsummary[:,0]/tempsummary[:,2],
```

we take the $0^{th}$ column, [:, 0], and divide those entries by the entries in the $2^{nd}$ column. [:, 2] (element-wise division). Then we replace the current $0^{th}$ column with the result of that division. Thus, the $0^{th}$ column now holds the average precipitation for each month. Similarly, line 32 replaces the $1^{st}$ column with the average temperature for each month. Finally, the summary matrix is displayed. The output of the program is given next.

**Output:**
```
0
1000000
2000000
3000000
4000000
5000000
6000000
7000000
8000000
9000000
[[     0.270      22.634 819888.000]
 [     0.171      23.114 749424.000]
 [     0.201      22.483 822384.000]
 [     0.110      21.651 797856.000]
 [     0.059      19.194 827304.000]
 [     0.050      18.303 811522.000]
 [     0.038      18.539 850433.000]
 [     0.030      19.508 860852.000]
 [     0.065      21.132 848951.000]
 [     0.129      22.176 795171.000]
 [     0.239      21.963 780197.000]
 [     0.277      22.995 815160.000]]

Process finished with exit code 0
```

From the output, we can see, for example, that the average hourly precipitation in January is 0.27 mm, and the average high temperature in May is 19.194 °C. We may wish to graph these values. To graph the monthly precipitation values, we could add the following lines:

```
m = np.arange(1,13,1)
plt.plot(m,tempsummary[:,0])
plt.xlabel('month')
plt.ylabel('precip')
plt.grid()
plt.show()
```

The graph is shown below.

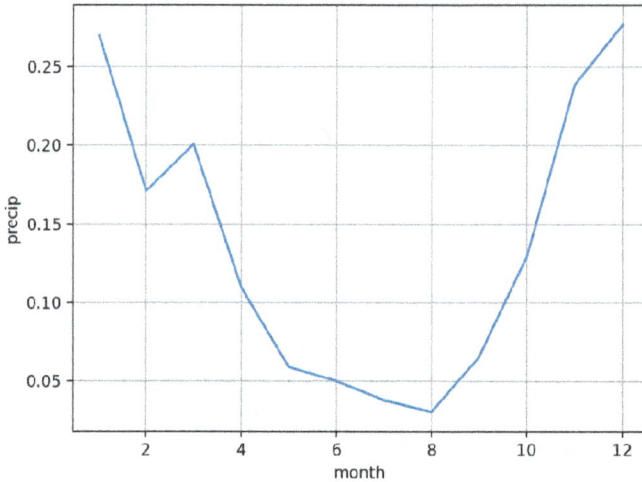

We could do a similar plot for the temperature by month. Notice that to get our plots we had to process all of the *tempstudy.csv* file. This takes a long time. It took my MacBook about 28 seconds to process the file. If we think we are going to use the summary information about precipitation and temperature frequently, we could write another file that holds just that small amount of information. Then, accessing that file would save a lot of time. We could add the following code to write the summary information to a file.

```
tempsum_file = open('tempsumm.csv','w')
for i in range(12):
    summrec = str(tempsummary[i,0])+','+str(tempsummary[i,1])+',\
              '+str(tempsummary[i,2])+'\n'
    tempsum_file.write(summrec)
tempsum_file.close()
```

Now, if we want to produce the graphs, we could just open this new file to get the summary information. Using the summary file, we can produce the same graph as before, except it takes about 0.13 seconds as opposed to 28 seconds. So, if we know we want to use the summary data often, it is worth our time to produce the new file. The code to produce the graph follows.

**Code:**

```
1   import numpy as np
2   import matplotlib.pyplot as plt
3   np.set_printoptions(precision=3,suppress=1,floatmode='fixed')
4   tempsumm_file = open('tempsumm.csv','r')
5   # set up a matrix to hold what we need
6   # we need a row for each month and columns for
```

```
7   # total precip, total temperature, and number of observations
8   tempsummary = np.zeros((12,3))
9   # read each line and fill the matrix
10  for i in range(12):
11      summrec = tempsumm_file.readline()
12      summvec = summrec.split(',')
13      for j in range(3):
14          tempsummary[i,j] = summvec[j]
15  m = np.arange(1,13,1) # a list of month values
16  plt.plot(m,tempsummary[:,0])
17  plt.xlabel('month')
18  plt.ylabel('precip')
19  plt.grid()
20  plt.show()
```

Notice that, in line 14, we do not need to convert the string to a floating-point value. This is because the tempsummary matrix was initialized as a numpy matrix. This means that the entries of the matrix are already assumed to be numeric. Thus, the conversion from string to numeric value is automatic whenever possible. For completeness, we show two more graphs next. One is the graph of the temperature by month. The other is the graph of the precipitation versus temperature. The reader should try to reproduce these graphs themselves.

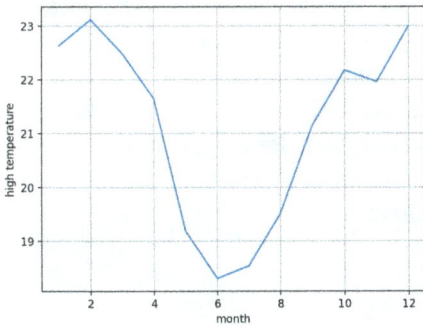

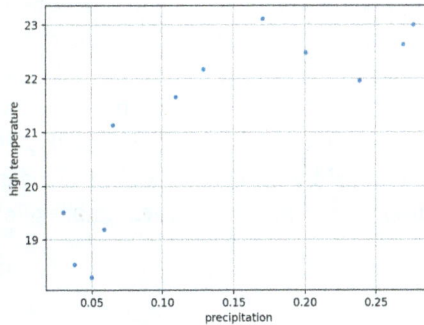

The right-hand graph indicates that higher precipitation values correspond to higher temperature values.

There are many more methods that help to handle files, and there are many more types of files that Python can handle. There are also packages available for Python, such as *Pandas*, that allow for efficient, sophisticated data management. The previous chapter just scratches the surface of the topic. But, it gives the reader an idea of the need to clean data and the need to be able to manipulate large files. While it is rare that we will deal with files with tens of millions of records, we will frequently deal with files with large amounts of data. We should be comfortable doing so.

**See Exercises 1–2.**

Now that we have some idea of how to acquire, clean, and process data, we turn our attention to describing the data with statistics.

## 6.2 Descriptive statistics

In this section, we briefly introduce some of the basic terms and ideas that are used when describing data. Entire courses are dedicated to this topic, so we do not attempt to offer a detailed presentation, but rather, we give a broad overview. Readers who find the topic interesting should pursue a course in statistics. We did some of this in the previous section when we found the average monthly precipitation and temperature. When we use such calculated values to describe data, we are using *descriptive statistics*. When we compute a number for an entire population, the number is called a *parameter*. When we compute the same number for a subset of the population (called a *sample*), then the number is called a *statistic*. For example, if we wanted to find the average height of students at a particular university, it may be possible to measure every student and find the average. In this case, we would have found the parameter that represents the mean height of the population of students. If, however, we take a random sample of students, and find the average height of the students in the sample, then we have found a statistic that represents the mean height. Furthermore, if we use the value from the sample to infer something about the population, then we are practicing *inferential statistics*. We will generally focus on sample statistics and inference because if you have access and the ability to measure the entire population, then no inference is needed. For a particular measurement (variable), we frequently want the mean of the measurement. This is the average that most people think of. If the variable is $x$, then, for a population, the mean is denoted by $\mu_x$. For a sample, the mean is denoted by $\bar{x}$. The mean is the sum of the observations divided by the number of observations.

$$\bar{x} = \frac{\text{sum of values}}{\text{number of values}}.$$

We can also compute what is called a *five-number summary* of the data. The five-number summary contains the following measurements.

- *Median:* The median, usually denoted by $M$, is the middle observation. If the number of data observations is even, it is the average of the middle two observations. If the observations are sorted from least to greatest, $x_1, x_2, \ldots, x_n$, then

$$M = \begin{cases} x_{\frac{(n+1)}{2}} & n \text{ is odd} \\ \frac{x_{\frac{n}{2}} + x_{\frac{n}{2}+1}}{2} & n \text{ is even} \end{cases}$$

- *Minimum or min:* This is the minimum value in the set of observations. If the observations are ordered, $min = x_1$.
- *Maximum or max:* This is the maximum value in the set of observations. If the observations are ordered, $max = x_n$.
- *First quartile:* This is the median of the observations that are at or below the median. We denote it by $Q_1$. It represents the value such that 25 % of the observations are at or below $Q_1$.
- *Third quartile:* This is the median of the observations that are at or above the median. We denote it by $Q_3$. It represents the value such that 75 % of the observations are at or below $Q_3$.

The mean and the median are measures of the center of the data. The *range* of the data is difference between the max and min. Likewise the *interquartile range* (IQR) is the difference between $Q_3$ and $Q_1$. The range and the IQR are measures of spread or dispersion, indicating how variable the observations are. Another very common measure of spread is the *standard deviation*, denoted by $\sigma$ (population) or $s$ (sample). The square of the standard deviation is called the *variance*. The formulas for variance are slightly different depending on whether we have data for the entire population or for a sample. Suppose we have $n$ observations. Then,

$$\sigma^2 = \frac{\sum_{i=1}^{n}(x_i - \mu)^2}{n} \text{ (population)} \quad \text{and} \quad s^2 = \frac{\sum_{i=1}^{n}(x_i - \bar{x})^2}{n-1} \text{ (sample).}$$

We can see from the formula that the variance measures the average squared distance from an observation to the mean. Hence, the larger the variance, the more spread out the data are.

Now, let's see how we might use some of these ideas. Let's return to our temperature study. Our first goal is to compute the five-number summary for the maximum temperature ($X$). To do this, we need to load the data into a matrix so that we can access the numeric values via the numpy commands. We could use code similar to that used in the previous section, but this is a good opportunity to show one of the many available methods that numpy provides to make our code more efficient. The numpy method `.genfromtxt` provides a way to load data from a text file into a numpy array. There are several arguments that allow us to select the data we want to convert to the type of value that we prefer. In the case at hand, we want the temperature values that are held in the 9th field (column) or the 8th index, and we know that the first line of the file contains headers. Then, to create the desired numpy array, we use the following code.

**Code:**

```
1   import numpy as np
2   np.set_printoptions(precision=3,suppress=1,floatmode='maxprec')
```

```
3   tempvals = np.genfromtxt('tempstudy.csv',dtype=float,delimiter=',',\
4                           usecols=(8), skip_header=True)
5   print('The number of values in tempvals is {}.'.format(len(tempvals)))
```

**Output:**

The number of values in tempvals is 9779142.


Process finished with exit code 0

This method of importing the data is substantially faster than loading and parsing each record, but it does require a higher level of coding sophistication. The .genfromtxt command takes several arguments (there are other option arguments that we have not used). The first argument is always the filename to be used. In our case, this is the temperature data held in *tempstudy.csv*. The second argument indicates that numeric strings should be converted to floating-point numbers (dtype=float). The delimiter=',' argument tells the Python interpreter that the file is comma delimited. If other delimiters are used, then they should be specified. The argument usecols=(8) indicates that we wish to load the data in position 8 ($9^{th}$ column). We could specify multiple columns by entering them as a comma-separated list within the parentheses. Finally, since we know the first line contains headers, we can instruct the method to skip the first line with skip_header=True. The print statement is included as a check so that we know data has actually been imported into the tempvals array. The output indicates that there are 9,779,142 values in the array, so it appears that the data have been imported. As a further check, we might print out the first 20 or so elements of tempvals to make sure they agree with the data file.

Now that we have the desired data, we can use numpy methods to compute the five-number summary. The following code is added to accomplish this.


**Code:**
```
6    M = np.median(tempvals)
7    Q1 = np.quantile(tempvals, .25)
8    Q3 = np.quantile(tempvals, .75)
9    min = np.min(tempvals)
10   max = np.max(tempvals)
11   R = max - min
12   IQR = Q3-Q1
13   print('Minimum = {:.2f}'.format(min))
14   print('Q1      = {:.2f}'.format(Q1))
15   print('Median  = {:.2f}'.format(M))
16   print('Q3      = {:.2f}'.format(Q3))
17   print('Maximum = {:.2f}'.format(max))
```

```
18   print('Range   = {:.2f}'.format(R))
19   print('IQR     = {:.2f}'.format(IQR))
```

**Output:**
```
The number of values in tempvals is 9779142.
Minimum = -3.20
Q1        = 18.20
Median  = 21.90
Q3        = 25.80
Maximum = 45.00
Range   = 48.20
IQR     = 7.60


Process finished with exit code 0
```

Note that we are assuming that the data has been cleaned prior to using these commands. Lines 6–10 compute the five-number summary while lines 11–12 compute the range and IQR respectively. We can tell by looking at the IQR and the median that the data are roughly symmetric with respect to the median. The "upper" half of the data has about the same spread as the lower half. To compute the mean and standard deviation, we use .mean and .std commands.

```
20   # compute the mean and standard deviation
21   xbar = np.mean(tempvals)
22   print('Mean    = {:.2f}'.format(xbar))
23   sd = np.std(tempvals,ddof=1)
24   print('Std.Dev.= {:.2f}'.format(sd))
```

**Output:**
```
The number of values in tempvals is 9779142.
Minimum = -3.20
Q1        = 18.20
Median  = 21.90
Q3        = 25.80
Maximum = 45.00
Range   = 48.20
IQR     = 7.60
Mean    = 21.11
Std.Dev.= 7.55


Process finished with exit code 0
```

In the .std method just used, doff=1 (degrees of freedom = $n - $ `) indicates that we are seeking a sample standard deviation. We would set ddof=0 or not include the option if we wanted the population standard deviation. We see that the mean and median for temperature are nearly the same value. This indicates that the distribution is symmetric. We can verify this by creating a *relative histogram* for the data. A relative histogram for a set of data divides the data values into equal numeric ranges (called classes) and gives the proportion of observed values that fall within each class. Python easily creates these histograms within matplotlib. We simply add the following to our existing code:

```
25  plt.hist(tempvals, density=1,edgecolor="black")
26  plt.show()
```

This produces the following graph.

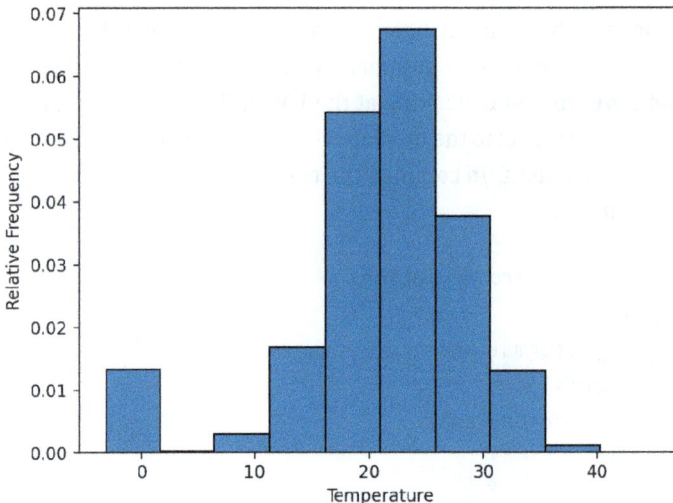

From the graph, we see that the days with temperatures at zero or below are separated from the rest of the data. This gives one cause to believe that the zero values for temperature are actually indicating that no value was reported. Thus, we should further clean this data by removing the days on which the temperature is zero or negative. The following histograms (Figure 6.2) show the shape of the distribution of temperatures when the errant days have been removed. It is clear from these graphs that the distribution is roughly symmetric.

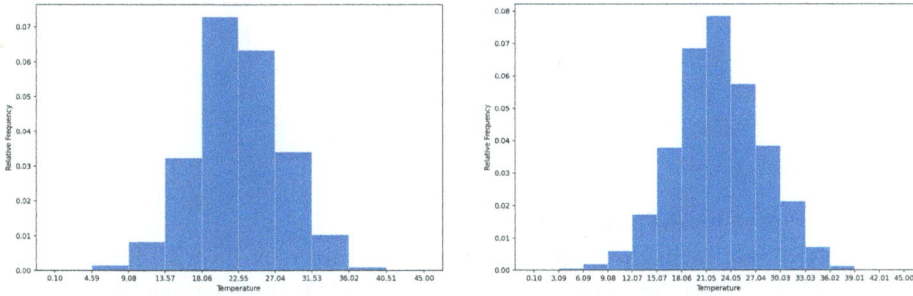

**Figure 6.2:** The histogram on the left uses 10 classes, while the histogram on the right uses 15 classes.

See Exercise 3.

## 6.3 Probability

Much of the power of statistics lies in the ability to estimate population parameters based on sample statistics. To do this properly, one must provide a level of confidence (or accuracy) for the estimate so that those who use the estimate have a sense of the risk assumed by using it. This measure of accuracy is often based on the likelihood of the sample that is being used. Such likelihood is called *probability*. We can think of the probability of an event as the long-term proportion of the number of times the event would occur in repeated trials of the experiment under consideration. The study of probability is both broad and deep, and there is no way to cover the subject with any detail in this text. Students are encouraged to take a course in probability and statistics if they have any interest in the study of random behavior or in the analysis of data in general. For our purposes, we will introduce the subject in a very general way and then proceed to apply some of the ideas that occur frequently in science.

As stated previously, the probability of an event $E$, denoted $P(E)$, is the long-term proportion of the occurrence of $E$. In some instances, the 'long-term' nature is hypothetical. For example, the probability of the winner of a sporting event is based on an imagining of repeated contests between the two teams. To compute probabilities for certain types of variables (continuous variables), we must know about the pattern of dispersion of the values of such a variable. From this pattern, we can define the *distribution* for the variable. There are several well-known, named distributions that are common: normal, Poisson, binomial, chi-square, student t, F, among others. The normal distribution is, by far, the most frequent distribution that is used (or assumed) when analyzing data. The curve that describes the likelihood of values (derived from the distribution) of a variable is called the *probability density function* (pdf) of the variable.

Let $f(x)$ be the probability density function for the random variable $X$. Then, $f(x)$ has the following properties:

$$(i)\ f(x) \geq 0 \quad \text{for all } x$$

$$(ii) \quad \int_{-\infty}^{\infty} f(x)\, dx = 1$$

$$(iii) \quad P(a \leq X \leq b) = \int_{a}^{b} f(x)\, dx,$$

where $P(E)$ represents the probability of $E$.

Suppose $X$ represents the height of students at a university. If we assume $X$ is normally distributed, then the density function, $f(x)$, would look something like the next graph.

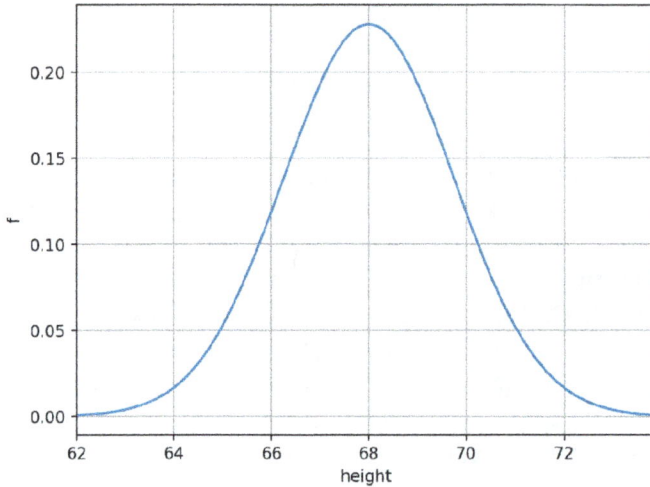

All normal density functions are bell-shaped curves with inflection points that are one standard deviation from the mean of the variable. The mean is the $x$ value that corresponds to the peak of $f(x)$.

When we take a sample of observations, then we generally do not know the mean of the population $\mu$. Hence, we use the mean of the sample, $\bar{x}$ as an estimate. Now, there are many theoretical details that we ignore here, but, for each sample of size $n$, we would likely get a different sample mean, $\bar{x}$. For example, suppose we take a sample of ten students and measure their heights. Further suppose that we compute the mean of the sample to be 69.2 in. Now, if we took a second sample of ten students and computed the mean height, the sample mean ($\bar{x}$) would likely not be 69.2 in. So, for the first sample, we would compute $\bar{x}_1$, for the second sample, we would get $\bar{x}_2$, and so on. Thus, the sample mean, $\bar{X}$, is itself a variable. When we study statistics and probability, we learn that the mean of the sample mean is the population mean, and the standard deviation of the sample mean is the population standard deviation divided by the square root of the sample size, i. e.,

$$\mu_{\bar{X}} = \mu$$

and

$$\sigma_{\bar{x}} = \frac{\sigma}{\sqrt{n}}.$$

For any normal distribution (or for large sample sizes), the variable $Z$ defined by

$$Z = \frac{x - \mu}{\sigma}$$

is normally distributed with a mean of 0 and a standard deviation of 1. $Z$ is called the standard normal distribution.

Finally, if the sample size is large, then the sample mean is approximately normally distributed, regardless of how the population is distributed. We have just put forth a lot of information very briefly, but the hope is that we have the basic tools needed to conduct some statistical analysis. From property (iii) of density functions, we see that probabilities are found using integration. Thus, a discussion of numerical integration seems appropriate at this point.

## 6.3.1 Numerical integration

Consider the function $f(x) = \frac{1}{\sqrt{2\pi}} e^{\frac{-x^2}{2}}$. The graph of the function on the interval $[-4, 4]$ is shown here.

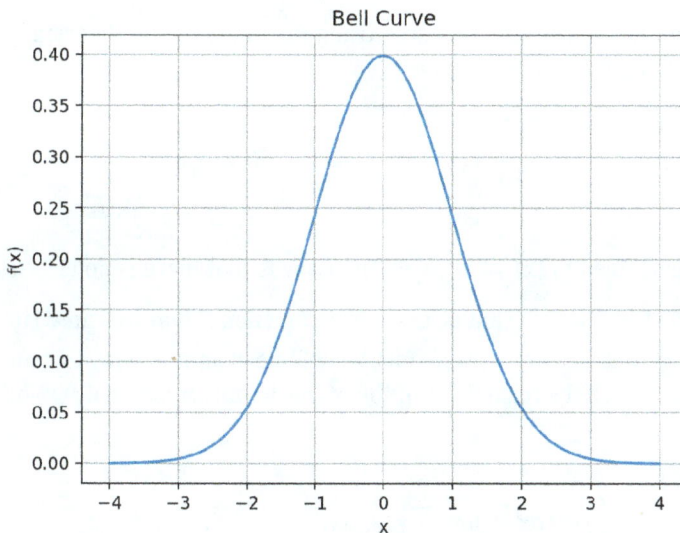

Bell Curve

The function $f(x)$ is the probability density curve for a normal variable $X$ (in this case, it is actually the standard normal variable $Z$ described previously) with a mean of zero and a standard deviation of 1. Thus, to find probabilities associated with $X$, we need to find the area under the this curve. For example, the probability that $X$ is between −2.3 and 1.2, denoted $P(-2.3 \leq X \leq 1.2)$, is represented by the area shown in the following figure.

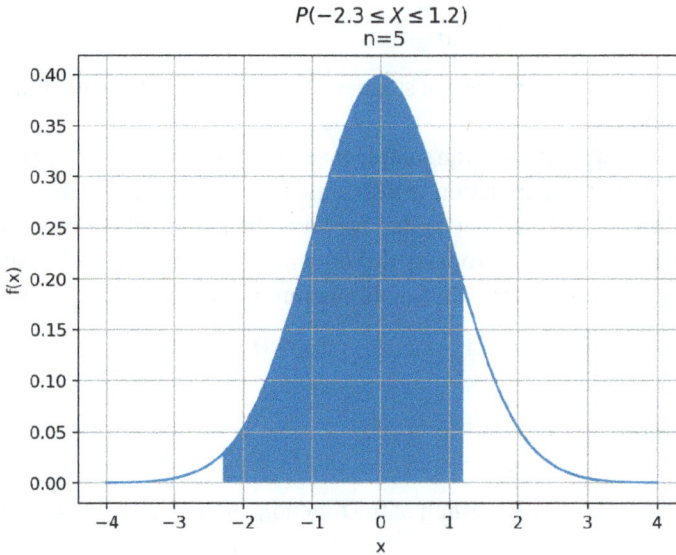

$P(-2.3 \leq X \leq 1.2)$
$n=5$

From calculus, we know that we can find this area using the fundamental theorem of calculus. Thus,

$$P(-2.3 \leq X \leq 1.2) = \int_{-2.3}^{1.2} f(x)\,dx = F(-2.3) - F(1.2),$$

where $F(x)$ is an antiderivative of $f(x) = \frac{1}{\sqrt{2\pi}}e^{\frac{-x^2}{2}}$. The issue is that there is no closed-form antiderivative of $f(x) = \frac{1}{\sqrt{2\pi}}e^{\frac{-x^2}{2}}$. That is, there is no function $F$ that we can write down in a nice, neat form so that $F'(x) = f(x)$. Thus, we will need another way to compute this definite integral. Recall the formal definition of the definite integral of $f(x)$ on $[a, b]$,

$$\int_{a}^{b} f(x)\,dx = \lim_{n \to \infty} \sum_{i=1}^{n} f(x_i)\Delta x.$$

Remember that this forms a partition of the interval $[a, b]$ using $n$ rectangles with the heights of the rectangles determined by $f(x)$. Then, we compute the area of each rectangle and sum the individual areas to approximate the integral. Allowing the number of rectangles, $n$, to go to infinity would then yield the true value of the integral. These summations of rectangular areas are called *Riemann sums*. The following figure gives the rectangular partitions for $n = 5$ and $n = 10$ for our probability example.

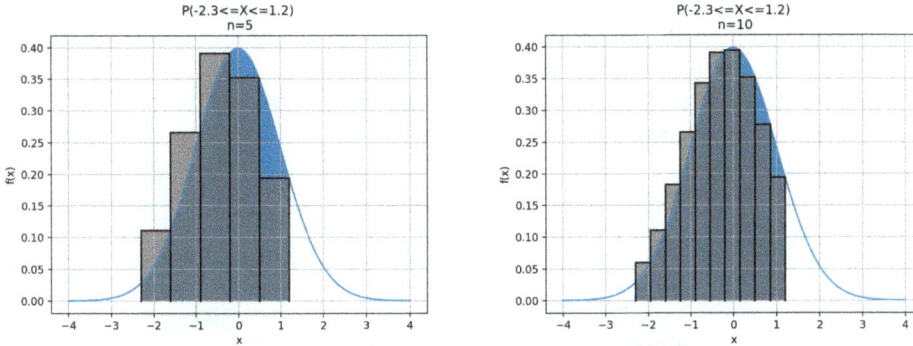

The areas that are either entirely gray or entirely blue indicate an error in the approximation. As the number of rectangles increases, this error decreases. The rectangles that are shown here use the $y$-value at the right endpoint of each subinterval as the height of the rectangle. Thus, these are called right Riemann sums. For our example, we have partitioned the interval $[-2.3, 1.2]$ into a set of $x$ values. For $n = 5$, each interval would need to be of width $\frac{1.2-(-2.3)}{5} = \frac{3.5}{5} = 0.7$. So, the $x$ values that define the partition are

$$x_0 = -2.3, \quad x_1 = -1.6, \quad x_2 = -0.9, \quad x_3 = -0.2, \quad x_4 = 0.5, \quad x_5 = 1.2,$$

and the corresponding $y$ values (found by substituting the $x$ values into $f(x)$) are

$$y_0 = 0.028, \quad y_1 = 0.111, \quad y_2 = 0.266, \quad y_3 = 0.391, \quad y_4 = 0.352, \quad y_5 = 0.194.$$

Thus, the area of each rectangle is given by

$$(\text{height of the rectangle})(\text{width of the rectangle}) = y_i \Delta x$$

for $i = 1, 2, 3, 4, 5$. We compute these areas and sum them to get

$$0.028(0.7) + 0.111(0.7) + 0.266(0.7) + 0.391(0.7) + 0.352(0.7) + 0.194(0.7) \approx 0.9200.$$

Because we are using right Riemann sums, $y_0$ is not used when calculating the areas. Since this is the right Riemann sum obtained using five rectangles, we denote this by $R_5 = 0.9200$. Similar calculations using 10, 50, and 100 rectangles yield

$$R_{10} = 0.90018, \quad R_{50} = 0.8799, \quad \text{and} \quad R_{100} = 0.8771.$$

Obviously, we want to use Python to produce these approximations. The code to do this is given next.

**Code:**

```
1   import numpy as np
2   import matplotlib.pyplot as plt
3   np.set_printoptions(precision=3,suppress=1,floatmode='maxprec')
4   # define the function
5   def stdnorm(x):
6       y = 1 / np.sqrt(2 * np.pi) * np.exp(-x ** 2 / 2)
7       return y
8
9   # get the number of rectangles
10  n = int(input('Enter the number of rectangles: '))
11  # limits of integration
12  a = -2.3
13  b = 1.2
14
15  # determine Delta x
16  dx = (b-a)/n
17  # create the partition of x values
18  x = np.arange(a,b+dx,dx)
19
20  # get the y values (heights of the rectangles
21  y = stdnorm(x)
22
23  # Compute the areas of each rectangle
24  A = y[1:]*dx
25  # Sum the areas
26  R = np.sum(A)
27  print('Riemann Sum is:',R)
```

**Output:**

```
Enter the number of rectangles: 5
Riemann Sum is: 0.9200101122108251
```

The areas are computed in line 24. Since we are using right sums, the first $y$ value is not used. Thus, we exclude it with y[1:] when multiplying the function values by $\Delta x = 0.7$. The sum of the areas is accomplished with line 26. The question that arises when using these sums to approximate the integration is: "How many rectangles do we need?" For our purposes, we will either choose a very large number of rectangles or use trial and error to see when the sum ceases to change substantially. For the current example, we would need thousands of rectangles to achieve the accuracy we would want when computing a probability. That is not terribly troubling since we have a computer that can do that easily. However, if we needed several such integrations, then it could begin to take appreciable time to get our results. Fortunately, there are other numerical integration techniques that are more accurate for fewer numbers of points in the partition.

**See Exercise 4.**

One modification to our numerical integration could be to change the shape used in each subinterval. So, instead of rectangles, we might use trapezoids as shown here.

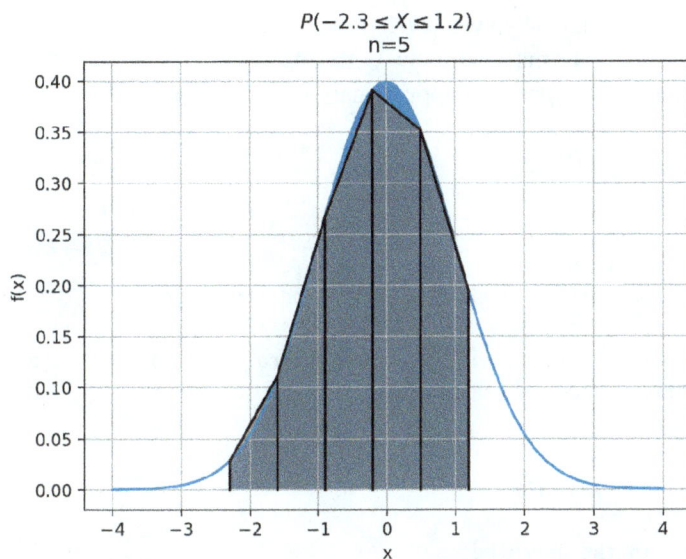

For a general trapezoid, the area is given by

$$A = \frac{1}{2}(b_1 + b_2)h,$$

where $b_1$ and $b_2$ are the bases (the lengths of the parallel sides) and $h$ is the height. In our case, the trapezoids are kind of on their sides. So the lengths of the bases are the function values at consecutive $x$ values, and the height of each trapezoid is the width of the interval, $\Delta x$. So the trapezoidal areas would be:

$$A_1 = \frac{1}{2}(y_0 + y_1)\Delta x$$

$$A_2 = \frac{1}{2}(y_1 + y_2)\Delta x$$

$$A_3 = \frac{1}{2}(y_2 + y_3)\Delta x$$

$$A_4 = \frac{1}{2}(y_3 + y_4)\Delta x$$

$$A_5 = \frac{1}{2}(y_4 + y_5)\Delta x.$$

Since $\Delta x$ and $\frac{1}{2}$ are common to all terms, we can factor these out of the sum to get an approximate total area (denoted by $T_5$) of

$$T_5 = \frac{\Delta x}{2}(y_0 + 2y_1 + 2y_2 + 2y_3 + 2y_4 + y_5).$$

We notice that all terms within the parentheses have a coefficient of 2 except the first and the last terms. This means that, when we develop code to do this sum, we will need to address the first and the last terms separately. With this example, we can write a Python function to compute the *trapezoidal* approximation to the integral. The code is given next.

**Code:**

```
1    import numpy as np
2    import matplotlib.pyplot as plt
3    np.set_printoptions(precision=3,suppress=1,floatmode='maxprec')
4
5    # use the trapezoidal rule to approximate the integral of f(x) from a to b
6    # we call the function traprule (trapezoidal rule)
7    def traprule(f,a,b,n):
8        dx = (b-a)/n
9        x = np.arange(a,b+dx,dx)
10       y = f(x)
11       # get the terms in the parentheses
12       # multiply all but the first and last y-values by 2
13       y[1:n] = 2*y[1:n]
14       # now multiply all terms by (delta x)/2
15       y = (dx/2)*y
16       # sum the areas
17       T = np.sum(y)
18       # return the value (T is the approximation to the integral)
19       return T
```

Note that, in line 13, the interior y values are replaced by twice the original y values, leaving the first and the last y values unchanged. This applies the proper coefficients to each of the y values. Then, line 15 multiplies each of these values by $\frac{\Delta x}{2}$. Finally, the sum is accomplished in line 17, stored in a local variable, T, and returned in line 19. Now, to use the `traprule` function, we simply need to call it with the appropriate arguments. For our example problem, we would add the following to our previous code.

**Code:**

```
20   # integrate the standard normal density function from -2.3 to 1.2
21   # using 50 subintervals.
22   IntegralVal = traprule(stdnorm,-2.3,1.2,50)
23   print('The approximate value of the integral is',IntegralVal)
```

**Output:**

```
The approximate value of the integral is 0.87408445763325
```

The standard normal density is well-known. Thus, we know the integral values for this function so that we can compare our results to the known results to see how well our method is performing. For this example, the integral should be approximately 0.87420622. Using trial and error for the amount of rectangles, we can show that it would take about 8,000 subintervals to achieve this amount of accuracy. If we attempt to achieve this amount of accuracy using our right-Riemann rule, we would need about 54,000,050 subintervals. Thus, the trapezoidal rule seams to be a great improvement with regard to the number of subintervals needed (and, hence, the amount of time required) to compute the integral. One more version of numerical integration that we see in calculus is *Simpson's* rule. Simpson's rule uses sets of three consecutive points to fit a parabola. Then, the integrals of all the parabolas are computed and summed to give the approximation to the overall integral. Figure 6.3 shows parabolas imposed on our example problem for $n = 6$.

The blue areas in the figure represent portions of the region of integration that are not accounted for by the coverage of the parabolas. We can see the parabolas provide far less error than either the Riemann sum or the trapezoid rule. The general form of Simpson's rule is stated as:

$$\int_a^b f(x)\,dx \approx \frac{\Delta x}{3}(f(x_0) + 4f(x_1) + 2f(x_2) + 4f(x_3) + 2f(x_4) + \cdots + 4f(x_{n-1}) + f(x_n)).$$

In order to use Simpson's rule, we must have an even number of subintervals so that we can fit the parabolas appropriately. If we apply Simpson's rule to the previous example, we need only about 80 subintervals to achieve the accuracy indicated previously. Thus, we see that Simpson's rule far out-performs the other methods.

**See Exercise 5.**

Python has other methods that are available for integration, and we can use those if desired. Now that we know how to do numerical integration, if we know the probability density function of a variable, we can compute probabilities by using our integration techniques.

$$P(-2.3 \leq X \leq 1.2)$$
$$n=6$$

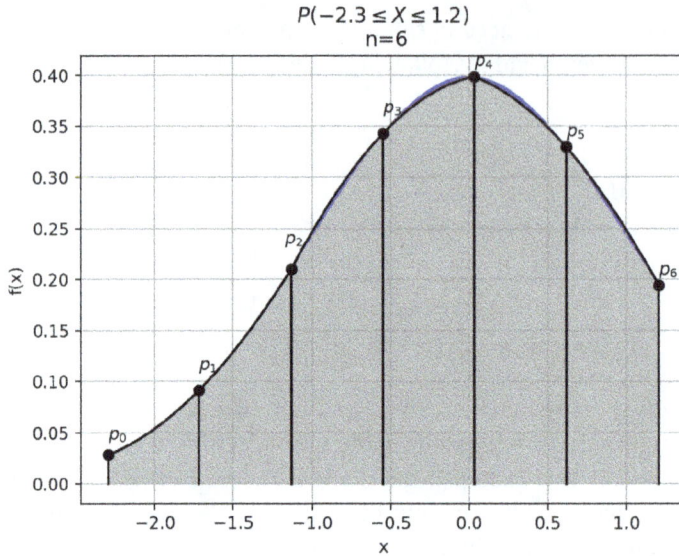

**Figure 6.3:** Multiple parabolas are fit to sets of three points: $(p_0, p_1, p_2)$, $(p_2, p_3, p_4)$, and $(p_4, p_5, p_6)$.

## 6.4 Confidence interval for the mean of a population

If we were to use the mean of a sample to represent the mean of a population (which is a common practice), then we are inferring the mean of the population from the sample. This action is part of what is called *inferential statistics*. When we make such inferences, we want to include some indication as to how accurate the estimate is. One way to do this is to construct what is called a *confidence interval*. A *P% confidence interval* for population parameter $\delta$ is an interval $(a, b)$ such that the probability, $p$, that $\delta$ is in $(a, b)$ is greater than or equal to $\frac{P}{100}$. For example, a 95 % confidence interval for the population mean, $\mu$, is an interval $(a, b)$ such that the probability of $a \leq \mu \leq b$ is greater than or equal to 0.95. Actually, the previous statement is subtly incorrect. The true statement is that there is $P$ % chance that the method will find an interval that contains the parameter. But, the general idea is that we are about $P$ % sure the interval contains the parameter.

So, suppose we take a sample of 100 students and find that the mean height of these students is 67.2 in. Further suppose that we do not know the mean height of all students (if we did, there would be no need to approximate it) and that we somehow know that the standard deviation of the population height is 1.75. We would like to construct a 95 % confidence interval for the mean height of the students in the university based on the sample we have taken. So, we want numbers $a$ and $b$ so that $P(a < \mu < b) = 0.95$. Based on the results given in Section 6.3, we know that $\mu_{\bar{x}} = \mu$ and that

$$\sigma_{\bar{x}} = \frac{\sigma}{\sqrt{n}} = \frac{1.75}{\sqrt{100}} = 0.175.$$

Knowing the standard deviation of the sample mean indicates that the shape of the density is known but the location of the center is not known since it would be centered at the population mean. Below are density curves for $\bar{x}$ that correspond to various values of the mean height of the population. The standard deviation associated with each curve is the same.

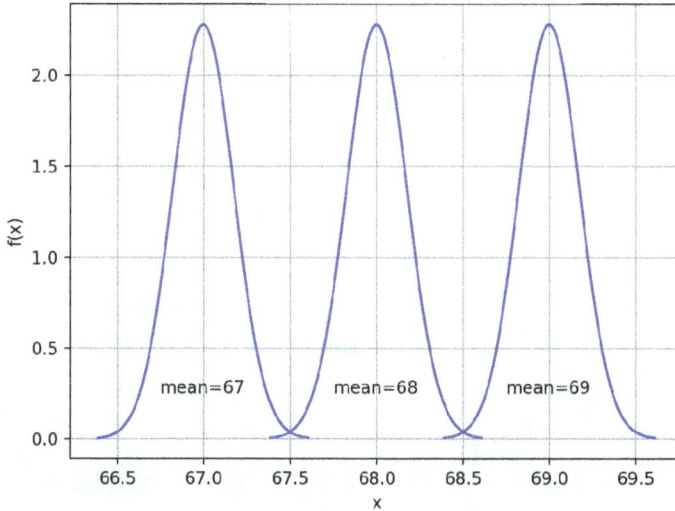

We transform to the standard normal to shift the mean to be 0. So, let

$$Z = \frac{\bar{x} - \mu}{\frac{\sigma}{\sqrt{n}}}.$$

Then, if we can find numbers $c$ and $d$ so that $P(c < Z < d) = 0.95$, we can transform $c$ and $d$ to determine the numbers that work for the original variable $X$ (height), i. e.,

$$P(c < z < d) = P\left(c < \frac{\bar{x} - \mu}{\frac{\sigma}{\sqrt{n}}} < d\right)$$

$$= P\left(c\frac{\sigma}{\sqrt{n}} < \bar{x} - \mu < d\frac{\sigma}{\sqrt{n}}\right)$$

$$= P\left(-\bar{x} + c\frac{\sigma}{\sqrt{n}} < -\mu < -\bar{x} + d\frac{\sigma}{\sqrt{n}}\right)$$

$$= P\left(\bar{x} - c\frac{\sigma}{\sqrt{n}} > \mu > \bar{x} - d\frac{\sigma}{\sqrt{n}}\right)$$

$$= P\left(\bar{x} - d\frac{\sigma}{\sqrt{n}} < \mu < \bar{x} - c\frac{\sigma}{\sqrt{n}}\right).$$

Thus, $a = \bar{x} - d\frac{\sigma}{\sqrt{n}}$ and $b = \bar{x} - c\frac{\sigma}{\sqrt{n}}$.

Since we know the mean and standard deviation of $Z$, we can find $c$ and $d$ by using something called the *inverse normal*. For $Z$, the density curve will look like this.

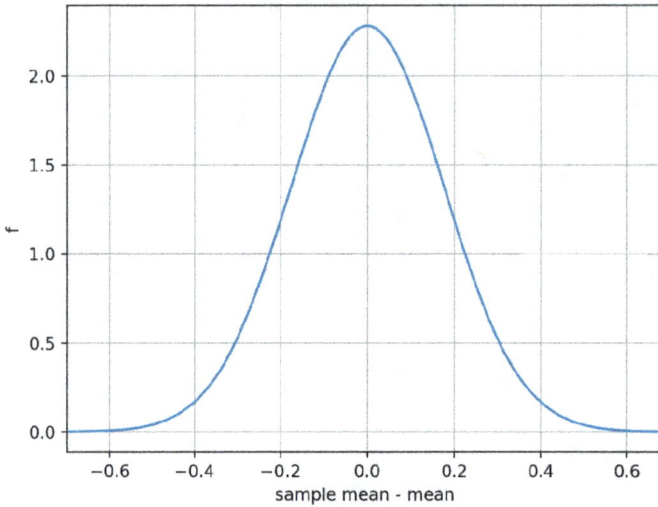

Thus, the goal is to find $c$ and $d$ such that the area under the curve between $c$ and $d$ is 0.95 as shown in the following.

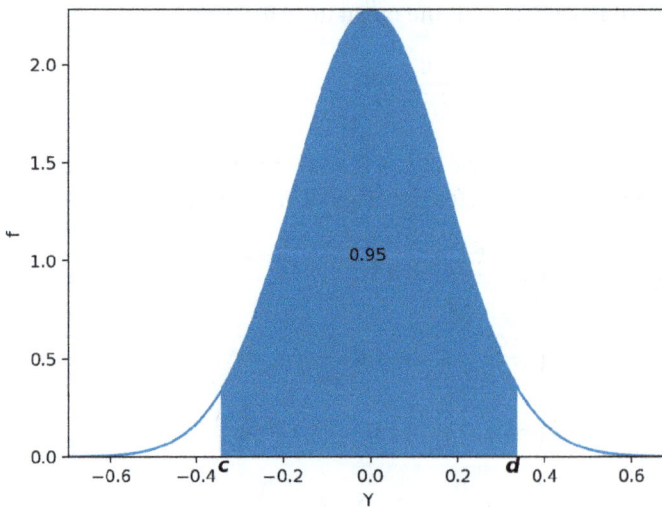

For a distribution, the inverse of the distribution is defined as

$$inverse(A) = a \quad \text{such that } P(x \le a) = A.$$

For example, *InvNormal*(0.8) = *a* means that $P(x \le a) = 0.8$. Thus, $d = InvNormal(0.975)$ and $c = InvNormal(0.025)$. In our example, we note that, if the area under the curve between *c* and *d* is 0.95, then there is an area of $1 - 0.95 = 0.05$ outside of the desired area. We will choose *c* and *d* such that the area below *c* is equal to the area above *d*. Thus, there is $\frac{0.05}{2} = 0.025$ in each tail of the distribution. Finally, since the inverse function uses the area below *d*, we combine the area below *c*, 0.025, and the shaded area, 0.95, to get a total area of 0.975 below *d*. To generate these numbers using Python, we need another package. We import the *stats* module from a package called *scipy*. The code is as follows.

**Code:**

```
1   import scipy.stats as stats
2   c = stats.norm.ppf(q=0.025,loc=0,scale=1.0)
3   d = stats.norm.ppf(q=0.975,loc=0,scale=1.0)
4   print('c =',c)
5   print('d =',d)
```

**Output:**

```
c= -1.9599639845400545
d= 1.959963984540054

Process finished with exit code 0
```

Line 1 imports the new package *scipy*. If this is not installed, the student will have to install this in the same way other packages have been installed. If the PyCharm editor is used, we can install new packages from within the editor. A Google search on something like "add scipy package in PyCharm" will lead to the required information. Line 2 finds the desired *c* value by using the *norm.ppf* command. This is the point percentile function that is equivalent to the inverse function. It gives the data value which would be the percentile value indicated by the first argument, 0.025. The *loc* argument is the mean, and the *scale* argument is the standard deviation. Because the normal distribution is symmetric about the mean, *c* and *d* are opposites of each other. The value 1.96 is called the critical value of Z and is often denoted by $z_{0.025}$. Now, we know that the interval for Z is $(-1.96, 1.96)$. The Z variable was our height transformed to standard normal. So

$$a = \bar{x} - d\frac{\sigma}{\sqrt{n}}$$
$$= 67.2 - 1.96\frac{1.75}{\sqrt{100}}$$
$$= 67.2 - 1.96(0.175)$$
$$= 66.857.$$

Likewise,

$$b = \bar{x} - c\frac{\sigma}{\sqrt{n}}$$

$$= 67.2 + 1.96\frac{1.75}{\sqrt{100}}$$

$$= 67.2 + 1.96(0.175)$$

$$= 67.543.$$

Hence, there is a 95 % likelihood that the true mean of the population is in the interval (66.857, 67.543). This is our 95 % confidence interval for the mean. We accomplish this in Python as follows.

**Code:**

```
1   import scipy.stats as stats
2   Import numpy as np
3   c = stats.norm.ppf(q=0.025,loc=0,scale=1.0)
4   d = stats.norm.ppf(q=0.975,loc=0,scale=1.0)
5   print('c =',c)
6   print('d =',d)
7   xbar = 67.2
8   sig = 1.75
9   a = xbar - d*sig/np.sqrt(100)
10  b = xbar - c*sig/np.sqrt(100)
11  print('a =',a)
12  print('b =',b)
```

**Output:**

```
c = -1.9599639845400545
d = 1.959963984540054
a = 66.8570063027055
b = 67.5429936972945
```

In general, if the population standard deviation is known, then a $P$ % confidence interval for the sample mean of a sample of size $n$ is given by

$$\left(\bar{x} - z_a\frac{\sigma}{\sqrt{n}}, \bar{x} + z_a\frac{\sigma}{\sqrt{n}}\right),$$

where $z_a$ is the critical value of $Z$ associated with $P$ %.

**See Exercise 6.**

In addition to giving some measure of certainty about the estimates that are being used, the confidence interval can also answer questions regarding the mean itself. For example, suppose we hypothesize that the mean height of the university students is 68 inches. Then, we take a sample and obtain the results as reported previously. There is a 95 % likelihood that the interval contains the actual mean of the population. Thus, since 68 is not included in the 95 % confidence interval, it is highly unlikely that 68 is the mean of the population.

In the prior discussion, we assumed that we did not know the population mean but we did know the population standard deviation. In most cases, this is unlikely. It is more common that neither the mean nor the standard deviation for the population is known. When this happens, we must modify our method for finding confidence intervals. We will use the sample's standard deviation as an estimate for the population's standard deviation. However, by using this estimate of $\sigma$, we can no longer assume that the associated sampling distribution is normal. Instead, we move to another distribution called the *student t* distribution. So, rather than taking the inverse of the normal, we take the inverse of the student t distribution. Suppose the heights of the 100 students in the sample are given in the table below.

| | | | | | | | | | |
|---|---|---|---|---|---|---|---|---|---|
| 65.654 | 67.263 | 67.186 | 64.808 | 66.137 | 67.487 | 67.214 | 72.155 | 69.201 | 68.274 |
| 67.610 | 70.088 | 66.167 | 68.535 | 66.216 | 67.382 | 66.867 | 68.633 | 65.349 | 69.423 |
| 67.729 | 67.250 | 65.304 | 68.566 | 62.739 | 65.567 | 69.029 | 67.769 | 62.608 | 64.695 |
| 66.873 | 64.753 | 70.209 | 65.162 | 66.258 | 69.359 | 69.038 | 68.135 | 66.837 | 67.007 |
| 69.321 | 67.853 | 69.662 | 65.779 | 65.295 | 66.136 | 69.085 | 69.504 | 67.754 | 65.131 |
| 66.470 | 67.661 | 68.761 | 65.610 | 67.970 | 69.646 | 69.795 | 64.861 | 66.320 | 67.531 |
| 65.426 | 66.926 | 70.485 | 67.880 | 66.498 | 68.265 | 65.429 | 68.368 | 66.464 | 67.190 |
| 70.934 | 68.399 | 68.986 | 68.162 | 65.521 | 66.383 | 66.250 | 63.739 | 67.099 | 63.716 |
| 66.573 | 62.929 | 67.399 | 66.959 | 66.416 | 68.436 | 71.919 | 66.320 | 67.314 | 66.979 |
| 67.733 | 66.684 | 67.074 | 67.174 | 68.305 | 65.056 | 67.582 | 67.737 | 64.178 | 70.572 |

We can compute the mean and standard deviation of this sample with the following code.

**Code:**

```
1  import numpy as np
2  import scipy.stats as stats
3
4  x = np.array([65.654, 67.263, 67.186, 64.808, 66.137, 67.487, 67.214,
5    72.155, 69.201, 68.274, 67.610, 70.088, 66.167, 68.535, 66.216, 67.382,
6    66.867, 68.633, 65.349, 69.423, 67.729, 67.250, 65.304, 68.566, 62.739,
7    65.567, 69.029, 67.769, 62.608, 64.695, 66.873, 64.753, 70.209, 65.162,
8    66.258, 69.359, 69.038, 68.135, 66.837, 67.007, 69.321, 67.853, 69.662,
9    65.779, 65.295, 66.136, 69.085, 69.504, 67.754, 65.131, 66.470, 67.661,
10   68.761, 65.610, 67.970, 69.646, 69.795, 64.861, 66.320, 67.531, 65.426,
11   66.926, 70.485, 67.880, 66.498, 68.265, 65.429, 68.368, 66.464, 67.190,
12   70.934, 68.399, 68.986, 68.162, 65.521, 66.383, 66.250, 63.739, 67.099,
13   63.716, 66.573, 62.929, 67.399, 66.959, 66.416, 68.436, 71.919, 66.320,
```

```
14    67.314, 66.979, 67.733, 66.684, 67.074, 67.174, 68.305, 65.056, 67.582,
15    67.737, 64.178, 70.572])
16  n = len(x)
17  xbar = np.mean(x)
18  print('xbar =',xbar)
19  s = np.std(x,ddof=1)
20  print('s =',s)
```

**Output:**
```
xbar = 67.2214
s = 1.8736521674111353


Process finished with exit code 0
```

From here, we proceed as we did when $\sigma$ was known except:
- we replace $\sigma$ with $s$;
- we take the *ppf* of the $t$ distribution instead of the *normal* distribution.

The $t$ distribution is dependent on the number of observations in the sample, $n$. The value of $n - 1$ is called the *degrees of freedom*. It is necessary to determine both the density function for $t$ and the inverse function for $t$. We forego the discussion of degrees of freedom in this text, but the interested reader could find the discussion in the many good statistic textbooks. Now, analogously to our previous example, we seek $c$ and $d$ such that $P(c < t < d)$ and find that

$$a = \bar{x} - d\frac{s}{\sqrt{n}} \quad \text{and} \quad b = \bar{x} - c\frac{s}{\sqrt{n}}.$$

Then, we can find the confidence interval by appending the following code.

**Code:**
```
19  c = stats.t.ppf(0.025,n-1)
20  d = stats.t.ppf(0.975,n-1)
21  print('c=',c)
22  print('d=',d)
23  a = xbar - d*s/np.sqrt(n)
24  b = xbar - c*s/np.sqrt(n)
25  print('a =',a)
26  print('b =',b)
```

**Output:**
```
xbar = 67.2214
s = 1.8736521674111353
```

```
c= -1.9842169515086832
d= 1.9842169515086827
a = 66.84962676081919
b = 67.59317323918081
```

```
Process finished with exit code 0
```

Hence, the confidence interval for this sample when the population standard deviation, $\sigma$, is unknown is $(66.850, 67.593)$. Notice that the magnitude of $c$ in this case was larger than that when we knew $\sigma$. This means that the confidence interval is likely to be wider. This makes sense because we are now estimating another quantity that makes us less sure of our estimate for the mean. Finally, we should mention that the *scipy.stats* package has a built-in method for finding confidence intervals. For the previous example, the following code will find the same confidence interval as was previously found.

**Code:**

```
1   import numpy as np
2   import scipy.stats as stats
3
4   x = np.array([65.654, 67.263, 67.186, 64.808, 66.137, 67.487, 67.214,
5       72.155, 69.201, 68.274, 67.610, 70.088, 66.167, 68.535, 66.216, 67.382,
6       66.867, 68.633, 65.349, 69.423, 67.729, 67.250, 65.304, 68.566, 62.739,
7       65.567, 69.029, 67.769, 62.608, 64.695, 66.873, 64.753, 70.209, 65.162,
8       66.258, 69.359, 69.038, 68.135, 66.837, 67.007, 69.321, 67.853, 69.662,
9       65.779, 65.295, 66.136, 69.085, 69.504, 67.754, 65.131, 66.470, 67.661,
10      68.761, 65.610, 67.970, 69.646, 69.795, 64.861, 66.320, 67.531, 65.426,
11      66.926, 70.485, 67.880, 66.498, 68.265, 65.429, 68.368, 66.464, 67.190,
12      70.934, 68.399, 68.986, 68.162, 65.521, 66.383, 66.250, 63.739, 67.099,
13      63.716, 66.573, 62.929, 67.399, 66.959, 66.416, 68.436, 71.919, 66.320,
14      67.314, 66.979, 67.733, 66.684, 67.074, 67.174, 68.305, 65.056, 67.582,
15      67.737, 64.178, 70.572])
16
17  n = len(x)
18  xbar = np.mean(x)
19  s = np.std(x,ddof=1)
20  print('sample mean =',xbar)
21  print('sample standard deviation =',np.std(x,ddof=1))
22  print('standard dev of sample mean =',s/np.sqrt(n))
23  a,b = stats.t.interval(alpha=0.95, df=n-1, loc=xbar, scale=s/np.sqrt(n))
24  print('left end of interval =',a)
25  print('right end of interval =',b)
```

The command to compute the confidence interval is shown in line 22. Note that the student t distribution is indicated (because confidence intervals can be constructed using many different distributions). The arguments of the .interval are as follows:
- *alpha*: the confidence level as a decimal;
- *df*: the degrees of freedom (sample size minus one);
- *loc*: the mean (or center);
- *scale*: the standard deviation.

**See Exercise 7.**

## 6.5 Hypothesis testing

Closely related to confidence intervals is another analytical technique called *hypothesis testing*. Hypothesis tests are used to determine if there is evidence to indicate that an assumed parameter is incorrect. For example, suppose a college publishes that the average GPA of student athletes is 2.35 on a 4.0 scale. A sample of 20 athletes is chosen, and the GPA of each athlete is recorded.

Sample Athlete GPAs

| | | | | |
|---|---|---|---|---|
| 2.46 | 2.2 | 2.09 | 2.84 | 2.82 |
| 2.19 | 2.76 | 2.72 | 2.98 | 2.22 |
| 2.74 | 2.28 | 2.47 | 2.13 | 2.81 |
| 2.98 | 2.01 | 2.67 | 2.34 | 2.62 |

We use the following code to find the mean and standard deviation of the sample.

**Code:**
```
1  import numpy as np
2  import matplotlib.pyplot as plt
3  np.set_printoptions(precision=3,suppress=1,floatmode='maxprec')
4
5  x=np.array([2.46,2.2,2.09,2.84,2.82,2.19,2.76,2.72,2.98,2.22,
6  2.74,2.28,2.47,2.13,2.81,2.98,2.01,2.67,2.34,2.62])
7  xbar = np.mean(x)
8  s = np.std(x,ddof=1)
9  print('Sample mean of GPAs:      {:.3f}'.format(xbar))
10 print('Sample standard deviation: {:.3f}'.format(s))
```

**Output:**

```
Sample mean of GPAs:        2.516
Sample standard deviation: 0.314


Process finished with exit code 0
```

Remember that ddof=1 indicates that we are computing the standard deviation of a sample instead of the population (ddof=0). Now, if the mean of the population is actually 2.35, then the mean of most random samples will be fairly close to 2.35. Of course, different samples will have different sample means. The mean of our sample is 2.516. This seems relatively high compared to the stated population mean of 2.35. Is it higher than expected? And, if it is higher than expected, what does that tell us? We will answer these questions by means of probabilities. Let $X$ represent GPA. We need to compute the probability that a sample would have a mean of 2.516 or higher. That is, we need $P(\bar{X} \geq 2.516)$. We call this probability the $p$-value. So,

$$p = P(\bar{X} \geq 2.516).$$

If $p$ is very low, then one of two things has happened. Either the sample that was chosen is actually an unusual sample, or the assumption that the population mean $\mu$ is 2.35 is incorrect. Thus, if $p$ is low, we may choose to take another sample to see if our results are repeatable. However, if the sampling techniques were appropriate and there is reason to resist taking another sample (which is often the case), then we would conclude that the evidence suggests that the assumed mean is incorrect. The next questions is: "What is 'low' when comparing the $p$ value?" The answer is up to the researcher and should be determined before the probabilities are calculated. This threshold probability that is used to make our decisions is designated as $\alpha$ and is called the *significance level*. Let's continue with our example. In the previous section, we used confidence levels to measure the reliability of the results. The value of $\alpha$ is 1 minus the confidence level. Thus, analogous to a 95 % confidence level, we would have a significance level of 0.05. The assumption that $\mu = 2.35$ is called the *null hypothesis*, and we denote it as

$$H_0 : \mu = 2.35.$$

In general, when testing means, the null hypothesis is usually in the form

$$H_0 : \mu = \mu_0.$$

In our example, $\mu_0 = 2.35$, so the null hypothesis is $H_0 : \mu = 2.35$. Given the null hypothesis, we might consider three possibilities:
- The actual mean is higher than $\mu_0$;
- The actual mean is lower than $\mu_0$;
- The actual mean is different than $\mu_0$.

Each of these options represents what we call an *alternative hypothesis*. The corresponding notations for our example are:

$$H_a : \mu > \mu_0$$
$$H_a : \mu < \mu_0$$
$$H_a : \mu \neq \mu_0.$$

Ideally, the alternative hypothesis should be determined before any statistics are computed, but it is often the case that the value of the sample statistic will dictate the alternative hypothesis. In our example, since $\bar{x} = 2.516$ is greater than the $\mu_0$ value of 2.35, we choose the alternative hypothesis to be $H_a : \mu > 2.35$. That is, we are testing the hypothesis that the mean GPA of student athletes is actually higher than reported. Common choices for $\alpha$ are 0.01, 0.05, and 0.10, but other values may be chosen. For our example, let's choose $\alpha = 0.05$. Thus, our hypothesis test can be expressed by

$$H_0 : \mu = 2.35$$
$$H_a : \mu > 2.35$$
$$\text{Significance} : \alpha = 0.05.$$

Now, the problem reduces to finding the probability that the sample mean is greater than or equal to the observed mean for our sample, that is, we need to find $P(\bar{X} \geq 2.516)$. To compute this probability, we proceed as we did in the previous section. If the sample size is large enough (and we know the population standard deviation), then we can use the normal distribution. If the sample size is small (say, less than 30) or the population standard deviation is unknown, then we use the $t$ distribution. In our example, the sample size is 20, and the population standard deviation ($\sigma$) is not known so we will use the $t$ distribution. The most common case, by far, is to use the $t$ distribution, and the corresponding hypothesis test is called a $t$ *test*. Since we are taking one sample (not comparing results of two samples), this is called a *one-sample t test*. Thus, we want to find the area under the $t$ curve that is to the right of $\bar{x} = 2.516$. See Figure 6.4.

To find this area, we need to find

$$\int_{2.516}^{\infty} f(x)\, dx,$$

where $f(x)$ is the density function shown. The formula for $f(x)$ (the density function for the $t$ distribution) is very complex. Fortunately, Python can compute this integral for us using something called the *cumulative distribution function* (cdf). The cumulative distribution function for a random variable $X$ is a function $F(x)$ such that

$$F(x) = P(X \leq x).$$

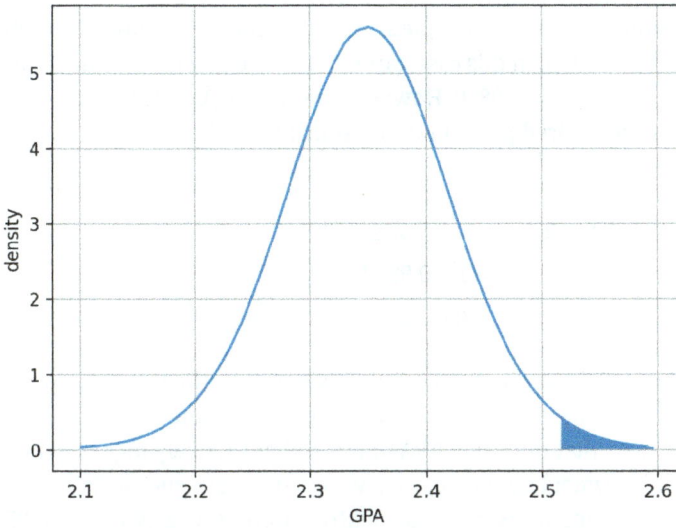

**Figure 6.4:** The shaded area represents $P(\bar{x} \geq 2.516)$.

For example, if $G(t)$ is the cdf for the sample mean GPA, then $G(2.516)$ is equal to the probability of the sample mean GPA being less than or equal to 2.516, i. e.,

$$G(2.516) = P(\bar{X} \leq 2.516).$$

Python can compute this value for us by adding the following to our previous code.

**Code:**
```
11   n = len(x)   #number of observations in the list
12   # get the cumulative distribution for xbar
13   cdf = stats.t.cdf(xbar,df=n-1,loc=2.35,scale=s/np.sqrt(n))
14   print('cdf(2.516) = {:.4f}'.format(cdf))
```

**Output:**
```
Sample mean of GPAs:        2.516
Sample standard deviation: 0.314
P(Xbar <= 2.516) = 0.9858

Process finished with exit code 0
```

The value of $P(\bar{X} \leq 2.516)$ is computed in line 13. The distribution function for $t$ needs to know the degrees of freedom, which is the size of the sample minus one. The degrees of freedom is included as the second argument of stats.t.cdf. Then, we must include the mean (loc) and the standard deviation (scale) that, as before, is the sample standard deviation divided by the square root of the sample size. And, of course, we have to supply the $x$ value that we are interested in. In this case, our value of interest is $\bar{x} = 2.516$. From the output we see that $P(\bar{X} \le 2.516) = 0.9858$. However, we want $P(\bar{X} \ge 2.516)$. Since the total area is 1, we can get the desired probability by subtracting the cdf value from 1. Thus,

$$P(\bar{X} \ge 2.516) = 1 - P(\bar{X} \le 2.516)$$
$$= 1 - 0.9898$$
$$= 0.0142.$$

So, the $p$ value for this test is 0.0142. Since the $p$ value is less than the threshold probability, $\alpha = 0.05$, we claim that there is sufficient evidence to reject the null hypothesis that $\mu = 2.35$. Thus, we suspect that the mean GPA of the athletes is likely higher than is being reported. If the $p$ value is not less than $\alpha$, then we cannot make such a claim.

We can also conduct the hypothesis test in a slightly different way. With $\alpha = 0.05$, we can find the sample mean, call it $c$, such that $P(\bar{X} \ge c) = 0.05$. This is similar to how we found the values for our confidence intervals. Because this is a "greater than" test, we need the inverse of $1 - 0.05 = 0.95$ to give us a *critical value* of the mean GPA.

**Code:**

```
1   import numpy as np
2   import matplotlib.pyplot as plt
3   np.set_printoptions(precision=3,suppress=1,floatmode='maxprec')
4   x=np.array([2.46,2.2,2.09,2.84,2.82,2.19,2.76,2.72,2.98,2.22,
5   2.74,2.28,2.47,2.13,2.81,2.98,2.01,2.67,2.34,2.62])
6   xbar = np.mean(x)
7   s = np.std(x,ddof=1)
8   print('Sample mean of GPAs:        {:.3f}'.format(xbar))
9   print('Sample standard deviation: {:.3f}'.format(s))
10  n = len(x)   #number of observations in the list
11  GPA_crit = stats.t.ppf(0.95,df=n-1,loc=2.35,scale=s/np.sqrt(n))
12  print('The critical value for GPA is {:.3f}.'.format(GPA_crit))
```

**Output:**

```
Sample mean of GPAs:        2.516
Sample standard deviation: 0.314
The critical value for GPA is 2.471.


Process finished with exit code 0
```

The critical GPA is found to be 2.471. The area to the right of this number is equal to $\alpha$ as shown here.

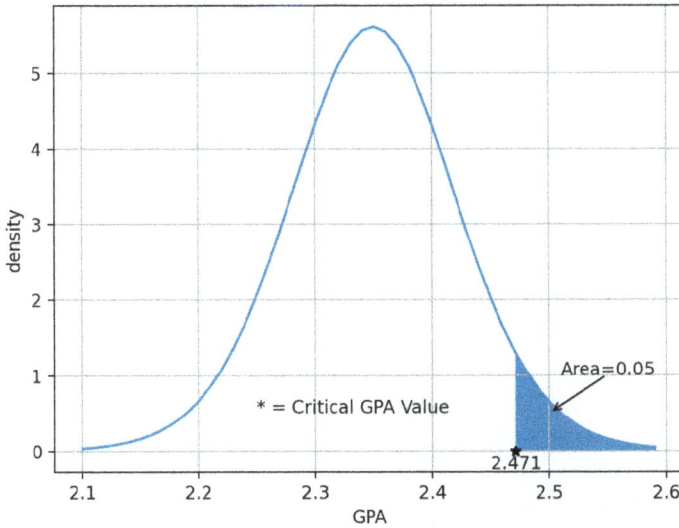

The shaded region on the graph is called the *critical region*. Now, the mean GPA for our sample is 2.516. Since 2.516 is to the right of 2.471, it lies within the critical region. This implies that the probability of obtaining a sample mean of 2.516 or higher must be less than $\alpha$. Thus, we can reject the null hypothesis. This gives us two perspectives of the hypothesis test:

- compute the $p$ value and compare it to $\alpha$;
- find the critical value and compare it to the sample statistic.

In our GPA example, we conducted a *one-tailed* test as seen by the critical region. In a general sense, if the alternative hypothesis is a 'greater than' statement, then the critical region is a region with area $\alpha$ located in the upper tail (right tail) of the distribution. If the alternative hypothesis is a 'less than' statement, then the critical region is a region with area $\alpha$ located in the lower tail (left tail) of the distribution, and if the alternative hypothesis is a 'not equal' statement, then the critical region is split into two regions, one in each tail having area $\frac{\alpha}{2}$. The corresponding critical regions and decision rules are shown in Figure 6.5.

When a two-tailed test is used, two $p$ values are theoretically considered, namely

$$P(\bar{X} \geq \text{observed sample mean})$$

and

$$P(\bar{X} \leq \text{observed sample mean}).$$

The $p$ value is the minimum of the two probabilities. However, only one of these probabilities needs to be computed because the other will be more than 0.5. If the observed

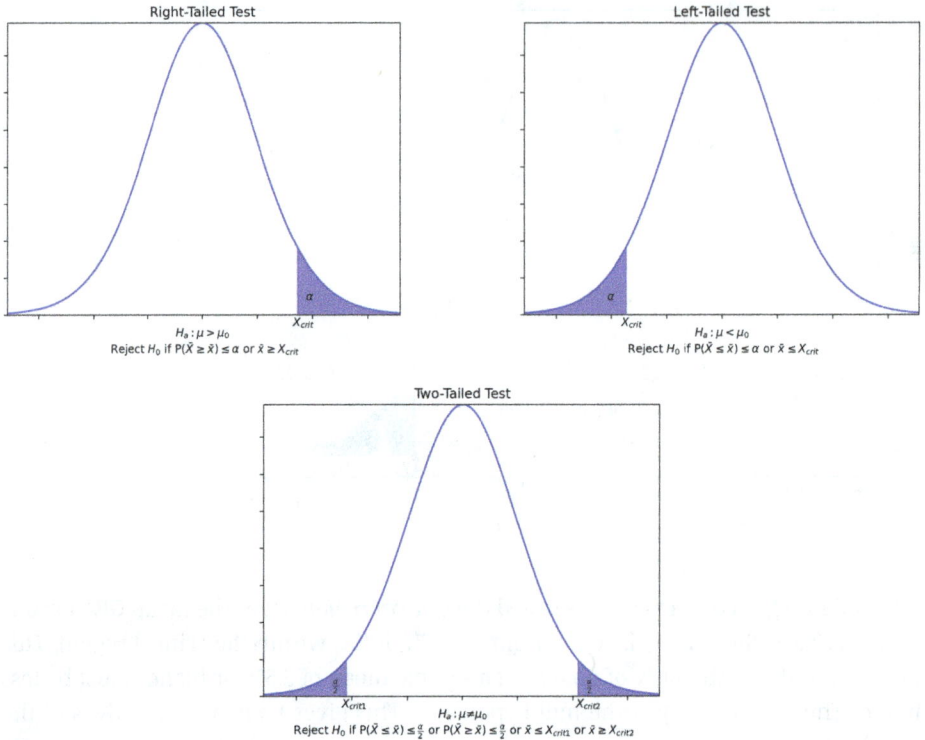

**Figure 6.5:** Critical regions and decision rules.

sample mean is greater than $\mu_0$, compute the greater-than probability. Otherwise, compute the lesser-than probability. For example, in our GPA scenario, if instead of $H_a > \mu_0$, we decided to test whether the mean GPA was simply not equal to the reported value. Then, the alternative hypothesis is $H_a \neq \mu_0$, and the test becomes a two-tailed test. Thus, having already computed the sample mean to be 2.516, we need to compute two probabilities,

$$P(\bar{X} \geq 2.516)$$

and

$$P(\bar{X} \leq 2.516).$$

We've already computed the first of these probabilities in our prior work, i. e.,

$$P(\bar{X} \geq 2.516) = 0.0142.$$

The probability associated with the less-than statement will then be

$$P(\bar{X} \leq 2.516) = 1 - 0.0142 = 0.9856.$$

Since the two probabilities must add up to be 1, computing one of them automatically gives us the other. The $p$ value for this test is, then, the same as the former test, 0.0142. However, because it is a two-tailed test, we must compare it to $\frac{\alpha}{2}$ instead of $\alpha$. Thus, if $\alpha = 0.05$, we compare $p$ to 0.025. Hence, in this case, we would again indicate that we should reject the null hypothesis.

See Exercise 8.

## 6.6 Comparing groups

The hypothesis tests in the previous section allowed us to infer whether a population parameter was reasonable, but we may also be faced with comparing the same parameter in different populations. For example, what if we wanted to know if the mean height of Americans was the same as the mean height of Europeans? This would require us to compare the two groups by using sample means. The technique used to compare groups depends, in part, on how many groups are involved.

### 6.6.1 Comparing means of two groups

Much like the previous section, we will us the $t$ distribution, but we must adjust the mean and standard deviation appropriately before computing $p$ values or critical values. Suppose we wish to compare the mean values of two groups to decide if they are significantly different. Then, the hypothesis test would be constructed as

$$H_0 : \mu_1 = \mu_2$$
$$H_a : \mu_1 \neq \mu_2,$$

where $\mu_1$ represents the mean of the population associated with the first group and $\mu_2$, the mean of the population of the second group. As before, our alternative test could also be $H_a : \mu_1 < \mu_2$ or $H_a : \mu_1 > \mu_2$ if the research question were posed in that manner. The general idea is to take a sample of values from each group, compute the sample means ($\bar{x}_1$ and $\bar{x}_2$) and standard deviations ($s_1$ and $s_2$), and use a $t$ test to determine if the means differ significantly. While we do not cover the theory in this text, it can be shown that the statistic

$$t = \frac{\bar{x}_1 - \bar{x}_2}{s_p \sqrt{1/n_1 + 1/n_2}}$$

follows the $t$ distribution with $n_1 + n_2 - 2$ degrees of freedom where $s_p = \sqrt{\frac{(n_1-1)s_1^2+(n_2-1)s_2^2}{n_1+n_2-2}}$. The quantity $s_p$ is called the *pooled standard deviation*. We can then use this statistic as we did in the previous section.

Let's consider an example. Suppose a study is done to evaluate a new diet plan. Two samples of ten people are chosen. One group (we will call them group A) eats as they normally do, while group B follows the diet plan. After ten weeks, the weight loss or gain is recorded for each individual with the following results.

Weight Gain/Loss in Lbs

| Group A | Group B |
|--------:|--------:|
| 2.61 | 0.62 |
| −4.29 | −4.03 |
| −2.5 | −5.32 |
| 2.34 | −6.92 |
| −4.31 | −0.84 |
| −4.37 | −4.88 |
| 2.01 | −7.83 |
| 0.16 | −4.84 |
| 1.04 | 2.07 |
| 3.13 | −4.88 |

Now, we use the $t$ distribution as we did before. Because this is a two-tailed test, we must compute two $p$ values and take the minimum of the two. Also, we either compare the $p$ value to $\frac{\alpha}{2}$, or we double the $p$ value and compare directly to $\alpha$. Most statistical packages do the latter. Thus, we will double the probability found. For our example, $\alpha$ is chosen to be 0.05. The code to accomplish this is given here.

**Code:**

```
1   import numpy as np
2   np.set_printoptions(precision=3,suppress=1,floatmode='maxprec')
3   import scipy.stats as stats
4
5   A = [2.61,-4.29,-2.5,2.34,-4.31,-4.37,2.01,0.16,1.04,3.13]
6   B = [0.62,-4.03,-5.32,-6.92,-0.84,-4.88,-7.83,-4.84,2.07,-4.88]
7   nA = len(A)
8   nB = len(B)
9   xbarA = np.mean(A)
10  sdA = np.std(A,ddof=1)
11  xbarB = np.mean(B)
12  sdB = np.std(B,ddof=1)
13  xdiff = xbarA - xbarB
14  pooledVar = ((nA-1)*sdA**2+(nB-1)*sdB**2)/(nA+nB-2)
15  sp = np.sqrt(pooledVar)
16  dofp = nA+nB-2
```

```
17   teststat = xdiff/(sp*np.sqrt(1/nA+1/nB))
18   print('test statistic:',teststat)
19   # mu_A < mu_B (group A lost more weight than B)
20   pvalL = stats.t.cdf(teststat,dofp)
21   # mu_A > mu_B (group A lost less weight than B)
22   pvalG = 1-stats.t.cdf(teststat,dofp)
23   pval = 2*np.min([pvalL,pvalG])
24   print('The p value is {:.3f}.'.format(pval))
```

**Output:**
```
test statistic: 2.296207907741645
The p value is 0.034.


Process finished with exit code 0
```

Lines 5–8 load the data and determine the number of observations in each group. The means and standard deviations for each group are computed in lines 9–12. The pooled standard deviation is computed in lines 14–15, and the test statistic is accomplished in line 17. Finally, the probabilities are computed. The "less-than" probability is computed in line 20, while the "greater-than" probability is computed in line 22. The $p$ value for this test is twice the minimum of these two probabilities. If the test were a one-tailed test, we would only need to find the appropriate probability. No doubling would be necessary. We see from the output that the $p$ value for this example is 0.034. Because $0.034 < 0.05$, we would reject the null hypothesis and indicate that there is evidence to do so.

The scipy package provides a method that will do all of this work with just one line of code. We could use the following.

**Code:**
```
print(stats.ttest_ind(a=A, b=B, equal_var=True))
```

**Output:**
```
Ttest_indResult(statistic=2.296207907741645, pvalue=0.033887154133051905)


Process finished with exit code 0
```

We see the same test statistic and $p$-value that we computed previously with much less work. However, the programmer must understand that the $p$ value represents a two-tailed value. Thus, if the alternate hypothesis is one-tailed, the correct interpretation must be made. For example, suppose we wish to test if the diet plan increased weight loss. Then, the alternative hypothesis in our example would be $H_a : \mu_1 > \mu_2$. We would use the exact same Python command, but the $p$ value would be half of the value that is returned. Thus, the $p$ value for the stated alternative hypothesis would be $p \approx 0.0169$. In general, the value of the sample means indicates the tail in which the $p$ value resides. For the general test, $H_0 : \mu_1 = \mu_2$, if $\bar{x}_1 < \bar{x}_2$, then the $p$ value derives from a left-tailed ("less-than") alternative test. If $H_a$ is contrary to this ($H_a : \mu_1 > \mu_2$), then there is no evidence to support rejection of the null hypothesis. If $\bar{x}_1 > \bar{x}_2$, the $p$ value derives from a right-tailed ("greater-than") alternative test.

**See Exercise 9.**

In many scenarios, an experiment may take multiple data measurements on the same sample of individuals. For example, in our diet scenario, instead of choosing different people for the groups (regular diet vs. plan diet), the researcher may select a sample of individuals to participate in the diet plan. Their weights are recorded before and after the plan period and then compared to determine if the diet works. Thus, the same individuals have two data values, before and after the diet. This is a reasonable design to test whether the plan is effective, but we must recognize that the two sets of numbers are no longer independent. To compare the mean weight before the diet to the mean weight after the diet, we use a *paired t test*. This paired character (dependence) affects the variance associated with the test, but, in Python, we need only make a small change to our method. Suppose the following data represents a pre-diet weight and a post-diet weight for each of ten individuals.

| Pre-Diet (lbs) | Post-Diet (lbs) |
|---|---|
| 152.61 | 149.62 |
| 145.71 | 144.97 |
| 147.5 | 143.68 |
| 152.34 | 142.08 |
| 145.69 | 148.16 |
| 145.63 | 144.12 |
| 152.01 | 141.17 |
| 150.16 | 144.16 |
| 151.04 | 151.07 |
| 153.13 | 144.12 |

Our hypothesis test would look like this:

$$H_0 : \mu_{\text{pre}} = \mu_{\text{post}}$$
$$H_a : \mu_{\text{pre}} > \mu_{\text{post}}.$$

The following code modifies the previous code to perform the correct $t$ test, but it also gives more guidance as to the computation of the $p$ value.

**Code:**

```
1   import numpy as np
2   np.set_printoptions(precision=3,suppress=1,floatmode='maxprec')
3   import scipy.stats as stats
4
5   A = [152.61, 145.71, 147.5,  152.34, 145.69,\
6         145.63, 152.01, 150.16, 151.04, 153.13]
7   B = [149.62, 144.97, 143.68, 142.08, 148.16,\
8         144.12, 141.17, 144.16, 151.07, 144.12]
9   xbarA = np.mean(A)
10  print('Pre-diet sample mean: {:.4f}'.format(xbarA))
11  xbarB = np.mean(B)
12  print('Post-diet sample mean: {:.4f}'.format(xbarB))
13  t,p = stats.ttest_rel(a=A, b=B)
14  print('t-stat = {:.4f}, p-value = {:.4f}'.format(t,p))
15  print('')
16  print('Assuming Ha: mu1 != mu2')
17  print('p value = {:.4f}'.format(p))
18  print('')
19  diffofmeans = xbarA - xbarB
20  if diffofmeans>0:
21      print('Assuming Ha: mu1 > mu2')
22      print('p value = {:.4f}'.format(0.5*p))
23  else:
24      print('Assuming Ha: mu1 < mu2')
25      print('p value = {:.4f}'.format(0.5*p))
```

**Output:**

```
Pre-diet sample mean: 149.5820
Post-diet sample mean: 145.3150
t-stat = 2.9341, p-value = 0.0166

Assuming Ha: mu1 != mu2
p value = 0.0166

Assuming Ha: mu1 > mu2
p value = 0.0083

Process finished with exit code 0
```

The code fills the sample data into arrays A and B in lines 5–8. So that correct decisions can be made later, the means of each sample are computed and displayed in lines 9–12.

Notice the call to the *t* test in line 13. Instead of `ttest_ind` which we used previously, we now use `ttest_rel` to conduct the paired *t* test. The method returns the test statistic and the two-tailed *p*-value and stores them in t and p. Lines 19–30 set up the logic to determine the likely *p*-value for the problem. If the difference of the sample means is positive, then the mean from group 1 was larger than the mean from group 2. Thus, the likely alternative hypothesis is the "greater-than" hypothesis. Otherwise, the likely alternative is the "less- than" hypothesis. In either case, the *p* value should be halved to represent a one-tailed test if that is actually the type of test that is being done.

There likely are Python packages that contain statistical methods that allow one to specify the type of alternative test, but as long as we know how to interpret results, the above *t* tests are suitable for independent or paired samples.

**See Exercise 10.**

## 6.6.2 Comparing means of more than two groups

To compare the means of more than two groups, we could do several tests between pairs of groups, but this can become cumbersome if there are several groups. If there are three groups to compare, say A, B, and C, then we would need to compare A to B, A to C, and B to C. That is not too bad. However, for four groups, A, B, C, and D, the number of pairwise comparisons grows to 6, and for more than four groups, the number of comparisons grows quickly. In addition to the number of pairwise tests that need to be done, the likelihood of an incorrect rejection of the null hypothesis increases with repeated pairwise testing. Thus, we seek a more efficient and effective way to compare group means. One commonly used technique is called the *analysis of variance*.

The method of analysis of variance (ANOVA) is generally used to test the equality of means for three or more groups. So, for three groups, the null hypothesis would be $H_0 : \mu_1 = \mu_2 = \mu_3$. The alternative hypothesis is that at least one of the population means is different from the others. When using ANOVA, the alternative hypothesis is always two-tailed. To justify the use of ANOVA, the following conditions should be met:
–   There are *n* independent samples (no paired data).
–   The associated populations have bell-shaped distributions.
–   The associated populations have the same variance.

While this may seem like a lot to ask, the ANOVA procedure is what one would call *robust*. This means that, if the data deviate from the above requirements slightly, the results of ANOVA will not change greatly and are still reliable. The ANOVA procedure actually uses variances to test the means of the various groups. Since the assumption is that the variance within each population is equal, two approximations to the variance are computed. One approximation involves the variation between groups, the other uses variations within each group. If the two approximations to the common population variance differ significantly, then the difference must be caused by differences in the means. Let's look a little closer. Suppose there are two study plans available for a student to help them prepare for a standardized test. Sixty students are divided into three groups: A, B, and C. Group A will study on their own, group B will use study plan 1, and group C will use study plan 2. Note that these groups are independent and that no one is in more than one group. The score for each individual is recorded and given in the following table.

| Group A | Group B | Group C |
|---------|---------|---------|
| 1026.0 | 1170.0 | 1248.0 |
| 989.0 | 1158.0 | 1301.0 |
| 961.0 | 1280.0 | 1492.0 |
| 1413.0 | 1388.0 | 1447.0 |
| 715.0 | 1174.0 | 1094.0 |
| 1013.0 | 957.0 | 1263.0 |
| 1130.0 | 1079.0 | 1462.0 |
| 1153.0 | 1154.0 | 1144.0 |
| 1029.0 | 1406.0 | 1133.0 |
| 1123.0 | 1125.0 | 995.0 |
| 1102.0 | 1228.0 | 1255.0 |
| 996.0 | 1090.0 | 1259.0 |
| 1103.0 | 1078.0 | 1182.0 |
| 1252.0 | 1123.0 | 1355.0 |
| 1109.0 | 828.0 | 1469.0 |
| 1296.0 | 1212.0 | 1158.0 |
| 911.0 | 1297.0 | 1339.0 |
| 1077.0 | 1246.0 | 1435.0 |
| 1180.0 | 1129.0 | 1311.0 |
| 1111.0 | 1316.0 | 1021.0 |

Our goal is to determine if the differing study plans produced significantly different scores. Thus, our null hypothesis is:

$$H_0 : \mu_A = \mu_B = \mu_C,$$

with an alternative of

$$H_a : \text{at least two means differ.}$$

The ANOVA technique assumes that each group comes from a population with mean $\mu$ and variance $\sigma^2$. The variability in the observations comes from two sources:
- variation within each group;
- variation between the groups

We use each of these sources to estimate the common variance $\sigma^2$. First, we consider the variance within each group. We can compute the sample variance $s^2$ for each group as we did earlier in this chapter. Thus, we can get $s_A^2$, $s_B^2$, and $s_C^2$. We compute a weighted sum of these variances to get what is called the *sum of squares due to error (SSE)*. So,

$$SSE = (n_A - 1)s_A^2 + (n_B - 1)s_B^2 + (n_C - 1)s_C^2,$$

where $n_A$, $n_B$, and $n_C$ are the number of observations in the respective groups. We then estimate $\sigma^2$ by

$$\frac{SSE}{n - k,}$$

where $n$ is the total number of observations and $k$ is the number of groups. So, for our example, we would have $\frac{SSE}{60-3}$. In general, the estimate just found is called the *mean square due to error, (MSE)*. Thus,

$$MSE = \frac{SSE}{n - k}.$$

Now we consider the variation between the groups. The variation between the groups is determined by a weighted sum of the squares of the differences between the means of each group and the overall mean. We call this the *sum of squares due to treatment (SST)*:

$$SST = n_A(\bar{x}_A - \bar{x}) + n_B(\bar{x}_B - \bar{x}) + n_C(\bar{x}_C - \bar{x})$$

To estimate $\sigma^2$, we divide the SST by one less than the number of groups. This is called the *mean square due to treatment (MST)*.

$$MST = \frac{SST}{k - 1}$$

So, *MSE* and *MST* are both estimates of $\sigma^2$. Thus, they should be nearly equal. If they are not, then our assumption of equal means is likely not true. Let's compute these estimates in Python.

**Code:**

```
1   import numpy as np
2
3   # load the data from the file
4   studydata = np.genfromtxt('studydata.csv', delimiter=',',skip_header=1)
5
6   # determine the number of rows and columns in the data
7   # each column is a group
```

```
8    # this code assumes that there are the same number of observations in
9    # each group
10   m,n = np.shape(studydata)
11
12   # compute the sample variance s^2 for each group
13   vars = np.zeros(n)
14   for i in range(n):
15       vars[i] = np.var(studydata[:, i], ddof=1)
16
17   # compute MSE
18   SSE = np.sum((m-1)*vars)
19   print('SSE: {:.4f}'.format(SSE))
20   MSE = SSE/(m*n-n)
21   print('MSE (within groups): {:.4f}'.format(MSE))
22
23   # compute MST
24   # get the means for each group
25   means = np.zeros(n)
26   for i in range(n):
27       means[i] = np.mean(studydata[:, i])
28
29   #get the overall mean
30   xbar = np.mean(studydata)
31
32   # compute MST
33   SST = np.sum(m*(means - xbar)**2)
34   print('SST: {:.4f}'.format(SST))
35   MST = SST/(n-1)
36   print('MST (between groups): {:.4f}'.format(MST))
```

**Output:**
```
SSE: 1189207.3000
MSE (within groups): 20863.2860
SST: 337715.0333
MST (between groups): 168857.5167

Process finished with exit code 0
```

The data is stored in the file *studydata.csv*. As was done before, the data is read from the file and loaded into a matrix using `.genfromtxt`. In this code, it is assumed that all of the treatment groups contain the same number of observations. It is possible to have a different number of observations in each group (as discussed previously), but the code would need modification to accommodate that circumstance. Line 10 determines the number of rows and columns in the data matrix. In this case, the number of rows is the number of observations in each group, and the number of columns is the number of treatment groups. Thus, m is the number of observations per group, and n is the number of groups. The sample variance for each group is computed using numpy.var with ddof=1 to indicate a sample rather than a population. Each group variance is stored within the vars vector. All of this is done in lines 13–15. To get *SSE*, we multiply each variance by one less than the number of observations in each group and sum the products. This is done in line 18. The total degrees of freedom (dof) is equal to the total number of observations minus the number of groups. Thus, $MSE = \frac{SSE}{mn-n}$ as is accomplished in line 20. At this point, we have *MSE* as the first estimate for $\sigma^2$. We then move to compute the mean square due to treatment (*MST*). This requires the mean for each group and the mean of all observations. The group means are stored in the vector denoted means. The group means and overall mean are computed in lines 25–30. The sum of squares due to treatment is computed in line 33, and *MST* is finally done in line 35. Helpful prints are interspersed to inform us of the results.

Now that the two estimates, *MSE* and *MST*, are known, we can compare them. Since they are meant to estimate the same parameter, if they are different, then one of our assumptions is likely incorrect. If the variances are actually equal, then the means of the groups must not all be the same. To make the comparison we compute a new statistic. The *F* statistic is given by

$$F = \frac{MST}{MSE}.$$

If *F* is large, then the variance due to the difference between the groups is larger than the differences within the groups. This would indicate that a difference in means is likely the cause of the difference in the estimates. The *F* values are governed by the *F* distribution. The *F* distribution needs two parameters, the degrees of freedom associated with the numerator (number of groups minus one) and the degrees of freedom associated with the denominator (total number of observations – number of groups). We use the .cdf method to determine the probability that *F* is less than the computed value. Then, we subtract that value from 1 to get the *p* value for the test. The Python code is given here.

**Code:**

```
37   # Compute the F statistic
38   F = MST/MSE
39   print('F statistic: {:.4f}'.format(F))
40
41   # compute the p-value
42   p = 1-stats.f.cdf(F,n-1,m*n-n)
43   print('p-value: {:.5f}'.format(p))
```

**Output:**
```
SSE: 1189207.3000
MSE (within groups): 20863.2860
SST: 337715.0333
MST (between groups): 168857.5167
F statistic: 8.0935
p-value: 0.00081
```

Because the *p*-value is very low, we would indicate that there is evidence to support the rejection of the null hypothesis, and, thus, infer that at least two of the group means are not equal. This result would likely motivate one to conduct pairwise tests at this point in order to determine which of the groups differ. There are also tests to determine if the variances are equal. There is much theory and detail that has been omitted in this discussion (like including a second level of treatment), but, hopefully, the idea of the test has been conveyed.

**See Exercise 11–12.**

The material presented in this chapter is just a small example of the many uses of probability and statistics. Inference can be made regarding several scenarios and quantities. While the material is compelling, our goal here is to expose you to the types of methods that are available. The hope is that the reader recognizes the power of such arguments, knows of their existence, and can seek further resources if needed.

## 6.7 Exercises

1. The *sci-kit* package (which we will use later) includes several example data sets. The data gives the values of several variables relating to tumors. The second attribute in the data represents whether the tumor is malignant (M) or benign (B). The other attributes are numeric measures of aspects of the tumor. Use the data in *wdbc-ex.csv* to write a new file that accomplishes the following:
   (a) Replace the strings (M or B) in the second field with zeros and ones such that 'M' is replaced with 1 and 'B' is replaced by 0 for each record.
   (b) Eliminate records that have missing data.
2. Modify the temperature summary program to find the average temperature and precipitation for each month of each year included in the data. Graph the results for temperature and precipitation in separate figures.
3. In the data reported in *wdbc.csv*, the third column gives the radius values for each tumor, the fourth column gives the texture of the tumor, and the seventh column gives the smoothness of each tumor.
   (a) Use the file-handling methods to load the necessary data into a numpy matrix of the proper size.

(b) Compute the five-number summary, the mean, and the sample standard deviation for radius, texture, and smoothness.

(c) Construct a relative frequency histogram for each of the variables using 20 classes in each case.

(d) Print the results in a neat and readable form.

4. (a) Modify the code in Section 6.3.1 so that the Riemann approximation to the integral is contained in a function called Rsum(fname, a,b,n) that takes as its arguments:
   – fname—name of the function to be integrated
   – a—lower limit of integration
   – b—upper limit of integration
   – n—number of rectangles (subintervals) to be used.
   The function should return the right Riemann approximation to the integral.

(b) Use the function to approximate

$$\int_0^3 (e^x - x^2)\, dx$$

with ten subintervals.

(c) Repeat the approximation using 1,000 subintervals, then 2,000 subintervals, and compare the results.

5. (a) Write a Python function to compute the Simpson's rule approximation to $\int_a^b f(x)\, dx$.

(b) Compute the following integral using the fundamental theorem of calculus. This will be the true value of the integral. Keep at least six decimal places.

$$\int_1^4 \left( \sin x - \frac{2}{x} \right) dx.$$

(c) Use Python to approximate the integral using right Riemann sums with 20 subintervals.

(d) Use Python to approximate the integral using the trapezoidal rule with 20 subintervals.

(e) Use Python to approximate the integral using Simpson's rule with 20 subintervals.

(f) Compute the absolute value of the error (|approximation − true value|) for each of the previous approximations. Comment on the results.

(g) Compare errors for the different approximations if 500 subintervals are used.

6. Suppose a university gives all students the same math placement test. A random sample of 20 students is chosen, and the sample mean is found to be 72. If the standard deviation of test scores is known to be 7, find a 90 % confidence interval for $\mu$.

7. In the *wdbc.csv* file, the third column contains the radius of each tumor. Load the radii into a numpy array, and find a 90 % confidence interval for the mean radius of a tumor.

8. A certain type of algae is studied and determined to grow at a particular rate when exposed to approximately eight hours of light per day. When studied under a microscope, it is found that the mean number of algae cells in a prescribed area is 45.6. A student decides to grow the same algae under conditions that provide 16 hours of light per day. The amount of algae for each observation in a sample is given in the following table.

| | | | |
|---|---|---|---|
| 62 | 59 | 51 | 58 |
| 67 | 47 | 56 | 67 |
| 45 | 74 | 66 | 59 |
| 59 | 66 | 58 | 55 |
| 54 | 58 | 58 | 64 |
| 53 | 69 | 68 | 68 |
| 60 | 50 | 56 | 50 |
| 53 | 68 | 75 | 50 |
| 47 | 62 | 64 | 68 |
| 46 | 58 | 46 | 39 |

Conduct a one-tailed hypothesis test with $\alpha = 0.01$ to determine if there is evidence to suggest that more light exposure increases algae growth.

9. The file *courserounds.csv* contains the scores for several golfers on each of four different courses. A golf organization claims that course 2 is more difficult than course 3. Does the data support this claim? Use the data and Python to compare the mean score for course 2 to the mean score for course 3 at a significance level of $\alpha = 0.05$. The test would be of the following form:

$$H_0 : \mu_{\text{course 2}} = \mu_{\text{course 3}}$$
$$H_a : \mu_{\text{course 2}} > \mu_{\text{course 3}}.$$

10. One hundred individuals volunteered to participate in a diet study. Their weights (in pounds) were recorded before the diet began and again ten weeks later when the diet period was over. The data are given in *weightdata.csv*. Conduct the appropriate hypothesis test (with $\alpha = 0.05$) to determine if the diet was effective at reducing a person's weight.

11. Suppose an investor is trying to determine if the various funds that are available have different rates of return. Each fund includes 50 stocks with varying rates for each stock. The data for four funds is given in the file *stockdata.csv* (this is not actual data). Conduct an ANOVA test to determine if there is evidence to reject the hypothesis that all the funds have the same rate of return.

12. (a) Modify the code given in the text so that the ANOVA method is a function with the argument of the function being a matrix of data observations. Redo Exercise 11 to verify that the function works correctly.

    (b) How might we modify the code to allow for differing numbers of observations for each group? Investigate the use of *args with Python functions.

# 7 Regression

In this portion of the book, we consider the idea of fitting a mathematical model to a set of data. A *mathematical model* is simply an equation or set of equations that forms a mathematical representation of a particular phenomenon. We begin with a straightforward, but incredibly useful, technique called *linear regression*.

## 7.1 Linear regression

We will use an example to illustrate the idea. Suppose we wish to study the number of births to mothers who are 15–17-years-old in the United States. Further, we conjecture that the number of such births is related to the poverty level of a community. A brief web search yielded a source (Mind On Statistics, 3rd edition, Utts and Heckard) that gives the data in the following, table with one row representing each of the 50 states and the District of Columbia. The data is stored in the file *poverty.txt*. The first column is the percentage of families in the state who live below the poverty level. The second column is the birth rate per 1,000 females who are 15–17-years-old.

| Location | PovPct | Brth15to17 | Brth18to19 | ViolCrime | TeenBrth |
|----------|--------|------------|------------|-----------|----------|
| Alabama | 20.1 | 31.5 | 88.7 | 11.2 | 54.5 |
| Alaska | 7.1 | 18.9 | 73.7 | 9.1 | 39.5 |
| Arizona | 16.1 | 35 | 102.5 | 10.4 | 61.2 |
| Arkansas | 14.9 | 31.6 | 101.7 | 10.4 | 59.9 |
| . | . | . | . | . | . |
| . | . | . | . | . | . |
| . | . | . | . | . | . |
| Wisconsin | 8.5 | 15.9 | 57.1 | 4.3 | 32.3 |
| Wyoming | 12.2 | 17.7 | 72.1 | 2.1 | 39.9 |

For our initial study, we wish to use the poverty percentage to predict the birth rates for mothers between 15–17-years-old. We let $X$ denote the poverty level and $Y$ the birth rate. In regression, the variable that is being predicted is called the *response variable*, while the variable that is used to do the prediction is called the *explanatory variable*. We begin by plotting all of the points given in the data. We want to plot only the points, not connected by lines. To do this, we build a list to hold the poverty percentages ($X$'s) and a list to hold the births ($Y$'s). We could type in each of the 51 values for each list, but, if we were using a large data set, this would be either impossible or highly impractical. This is one of the reasons that we studied the file operations in Section 6.1. So we will load the data directly from the file and store it employing formats that are convenient for us to use in Python. The code to do this follows.

https://doi.org/10.1515/9783110776645-007

**Code:**

```
1   import numpy
2   import matplotlib.pyplot as plt
3
4   #open the data file
5   pov_file = open('poverty.txt','r')
6   #see how many lines are in the file
7   count = len(pov_file.readlines())
8   pov_file.close()
9   #since the first line contains headers, there are one fewer
10  #lines of numeric data
11  count = count-1
12
13  #set up some storage space for the data
14  x = numpy.zeros((count,1))
15  y = numpy.zeros((count,1))
16  #now we will reopen the file and read it line by line
17  #the first line of this file is a header
18  pov_file = open('poverty.txt','r')
19  headers = pov_file.readline()
20  #I printed the headers just in case I wanted to reference them
21  print(headers)
22  #now read the rest of the file one line at a time
23  for i in range(count):
24      #get the next line
25      l = pov_file.readline()
26      #split the line into separate fields
27      fields = l.split()
28      #the second field is the poverty percent. this is our x value
29      x[i] = float(fields[1])
30      #the third field holds the births that we want.  store them in y
31      y[i] = float(fields[2])
32  #close the file
33  pov_file.close()
34  plt.plot(x,y,'.')
35  plt.xlabel('Poverty Percentage')
36  plt.ylabel('Births to 15 to 17 Year Old Mothers')
37  plt.grid()
38  plt.show()
```

The code is commented generously, and students are encouraged to read through the code to be sure they understand how the data is loaded and plotted. The plot of the data points is called a *scatter plot*. The scatter plot for our example is given next.

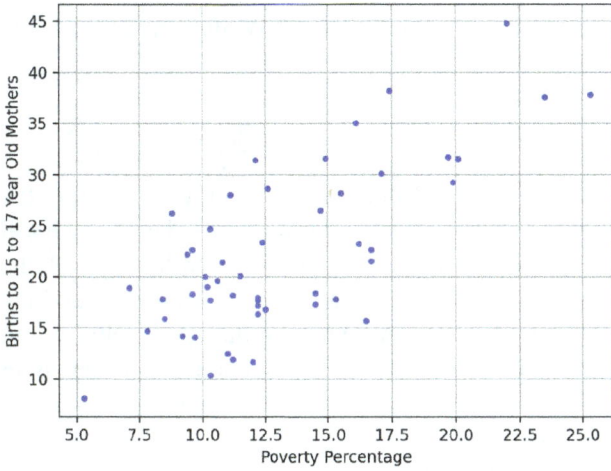

From the plot we can see a general trend that indicates the number of births increases as the poverty percentage increases. Also, notice that these points do not necessarily lie on a function. If we had two locations with the same poverty percentage, we would likely have different numbers of births for each of these locations. Hence, the points would not satisfy the vertical line test for functions. However, we may wish to predict the average number of births for a given poverty percentage, and a linear model seems appropriate in this case because there does not seem to be any indication that the points would form a "curve" as poverty percentage increases. We will address the appropriateness of the model soon, but, for now, we want to try to fit the best line through the data points. So, what do we mean by "best?" There are many lines that may appear to "fit" or represent the data well. See the following figure.

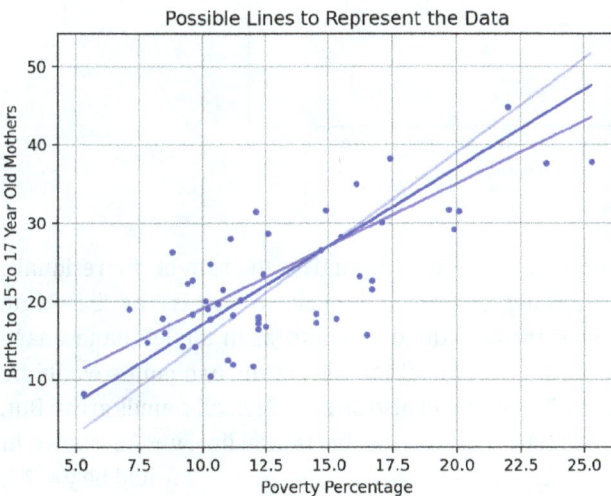

We need to define precisely what we mean by the "best line". We do so as follows. Suppose that the line that represents the data is given by

$$\hat{y} = mx + b.$$

We will use $y$ to represent the actual $y$ value of the data point and $\hat{y}$ to represent the prediction for $y$ given by the line. So, $\hat{y}$ is the prediction for $y$ associated with $x$. We will also denote our points as $(x_i, y_i)$, i.e., the first point in the data list will be $(x_1, y_1)$, the second will be $(x_2, y_2)$, and so on. So the actual $y$ value of the $n^{\text{th}}$ point is $y_n$, and the associated prediction is $\hat{y}_n = mx_n + b$. The difference between the actual value, $y$, and the predicted value, $\hat{y}$, is called the *residual*. Thus, for each data point $(x_i, y_i)$, there is an associated residual, $r_i$, where

$$r_i = y_i - \hat{y}_i.$$

The following graph illustrates a few of the residuals from our birth rate data for a particular linear model.

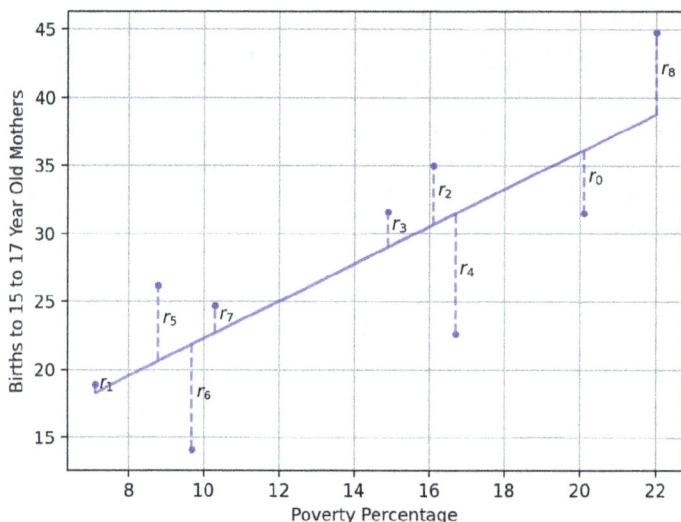

We might define the best line to be the line that minimizes the sum of the residuals. However, this is too simplistic. Consider a data set with just two points. The "best" line would clearly be the line that goes through the two points (as in the left-hand image shown in Figure 7.1). In such a case, the residual for each of the two points would be zero, and, hence, the sum of the residuals would also be zero. That all sounds great. But, suppose the points were $(2, 5)$ and $(4, 10)$. Consider the horizontal line, $y = 7.5$ (shown in the right-hand image of Figure 7.1). Then, the predicted value for $x = 2$ would be $\hat{y} = 7.5$, and the predicted value for $x = 4$ would be $\hat{y} = 7.5$. Thus, the residual for the first point is $r_1 = 5 - 7.5 = -2.5$ and, for the second point, $r_2 = 10 - 7.5 = 2.5$. So the sum of the residuals is $-2.5 + 2.5 = 0$.

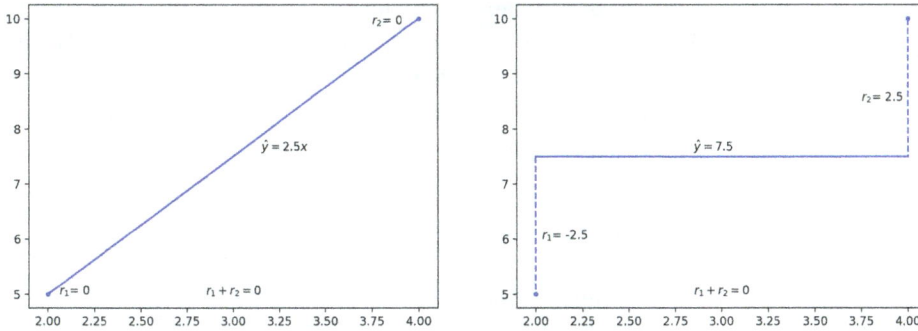

**Figure 7.1:** For both lines, the sum of the residuals is zero. However, the line on the left is clearly a better fit.

By measure of the sum of the residuals, both lines are equally "good." We know that the line through the points is better than the horizontal line because it predicts both points perfectly. So, we must refine our idea of "good" or "best" fit. We could revise the idea by minimizing the sum of the absolute values of the residuals. Hence, the new goal would be to minimize

$$F = \sum_{i=1}^{n} |r_i|.$$

This is a good idea, but absolute value functions tend to have corners which means that there are places where the function does not have a derivative. This makes the minimization process a bit more difficult. So, instead of using the absolute value of the residuals, we square each residual to make it positive and preserve the differentiability of the function. Finally, we define the *line of best fit* to be the line that minimizes

$$F = \sum_{i=1}^{n} (r_i)^2$$

where $n$ is the number of data points. Remember that $r_i = y_i - \hat{y}_i$, and, $\hat{y}_i = mx_i + b$. So $F$ can be written as

$$F = \sum_{i=1}^{n} (y_i - \hat{y}_i)^2$$
$$= \sum_{i=1}^{n} (y_i - (mx_i + b))^2$$
$$= \sum_{i=1}^{n} (y_i - mx_i - b)^2.$$

In $F$ as shown, all of the $x_i$'s and $y_i$'s are known. They are given in the $n$ points from our data. So, the unknowns are $m$ and $b$ that will define the regression line that we are seeking. Thus, $F$ is a function of $m$ and $b$, $F(m, b)$. If $F$ were a function of $x$, we would minimize it by finding the derivative and solving $F' = 0$ (remember your first calculus course?). We proceed very similarly when $F$ is a function of two variables. We need to take two different derivatives, one treating $m$ as the variable and one treating $b$ as the variable. We call these *partial derivatives*. When we treat $m$ as the variable, we think of $b$ as a constant. The partial derivative is denoted by $\frac{\partial F}{\partial m}$. So, the $x_i$'s and $y_i$'s are just numbers, and we treat $b$ as a constant to get

$$\frac{\partial F}{\partial m} = \sum_{i=1}^{n}[2(y_i - mx_i - b)(-x_i)].$$

Likewise, treating $m$ as a constant, we get

$$\frac{\partial F}{\partial b} = \sum_{i=1}^{n}[2(y_i - mx_i - b)(-1)].$$

Just as in the first semester of calculus, we must set the derivatives equal to zero and solve the resulting system.

$$\sum_{i=1}^{n}[2(y_i - mx_i - b)(-x_i)] = 0$$

$$\sum_{i=1}^{n}[2(y_i - mx_i - b)(-1)] = 0$$

Expanding the sums enables us to isolate the terms with $m$ and $b$.

$$\sum_{i=1}^{n}[2(y_i - mx_i - b)(-x_i)] = \sum_{i=1}^{n}(-2x_i y_i) + \sum_{i=1}^{n}(2x_i^2 m) + \sum_{i=1}^{n}(2x_i b)$$

$$= 2\sum_{i=1}^{n} -x_i y_i + 2m\sum_{i=1}^{n} x_i^2 + 2b\sum_{i=1}^{n} x_i$$

and

$$\sum_{i=1}^{n}[2(y_i - mx_i - b)(-1)] = \sum_{i=1}^{n}(-2y_i) + \sum_{i=1}^{n}(2x_i m) + \sum_{i=1}^{n}(2b)$$

$$= 2\sum_{i=1}^{n} -y_i + 2m\sum_{i=1}^{n} x_i + 2b\sum_{i=1}^{n} 1.$$

Thus, the system to be solved becomes

$$2\sum_{i=1}^{n} -x_i y_i + 2m\sum_{i=1}^{n} x_i^2 + 2b\sum_{i=1}^{n} x_i = 0$$

$$2\sum_{i=1}^{n} -y_i + 2m\sum_{i=1}^{n} x_i + 2b\sum_{i=1}^{n} 1 = 0.$$

Divide all terms by 2, and rearrange so that terms that do not include $m$ or $b$ are on the right-hand side of the equations. Then, we have

$$m\sum_{i=1}^{n} x_i^2 + b\sum_{i=1}^{n} x_i = \sum_{i=1}^{n} x_i y_i$$

$$m\sum_{i=1}^{n} x_i + b\sum_{i=1}^{n} 1 = \sum_{i=1}^{n} y_i.$$

This is a linear system of equations with unknowns $m$ and $b$. In matrix form, the system would be $AX = B$ where

$$A = \begin{bmatrix} \sum_{i=1}^{n} x_i^2 & \sum_{i=1}^{n} x_i \\ \sum_{i=1}^{n} x_i & \sum_{i=1}^{n} 1 \end{bmatrix}, \quad X = \begin{bmatrix} m \\ b \end{bmatrix}, \quad \text{and} \quad B = \begin{bmatrix} \sum_{i=1}^{n} x_i y_i \\ \sum_{i=1}^{n} y_i \end{bmatrix}.$$

We can now use Python to construct these matrices and solve the system as we did in Section 4.3. The complete code is given next.

**Code:**

```
1   import numpy as np
2   import matplotlib.pyplot as plt
3
4   #open the data file
5   pov_file = open('poverty.txt','r')
6   #see how many lines are in the file
7   count = len(pov_file.readlines())
8   pov_file.close()
9   #since the first line contains headers, there is one less actual
10  #lines of data
11  count = count-1
12
13  #set up some storage space for the data
14  x = np.zeros((count,1))
15  y = np.zeros((count,1))
16  #now we will reopen the file and read it line by line
17  #the first line of this file is a header
18  pov_file = open('poverty.txt','r')
19  headers = pov_file.readline()
20  #i printed the headers just in case i wanted to reference them
21  print(headers)
22  #now read the rest of the file and store the x's and the y's
23  for i in range(count):
24      #get the next line and store it in l
```

```
25    l = pov_file.readline()
26    #split the line into separate fields (assumes space delimited)
27    fields = l.split()
28    #the second field (which will have an index of 1) is the poverty percent.
29    #this is our x value
30    x[i] = float(fields[1])
31    #the third field holds the births that we want.  store them in y
32    y[i] = float(fields[2])
33 #close the file
34 pov_file.close()
35
36 #our variables are m and b.  we need the matrix of coefficients
37 A = np.zeros((2,2))
38 #first row of coefficients
39 A[0,0] = np.sum(x*x)
40 A[0,1] = np.sum(x)
41 #second row of coefficients
42 A[1,0] = np.sum(x)
43 #the sum of 1 is equal the number of terms in the sum
44 A[1,1] = len(x)
45 #now we need the right hand side
46 B = np.zeros(2)
47 B[0] = np.sum(x*y)
48 B[1] = np.sum(y)
49 print('A=',A)
50 print('B=',B)
51 #now solve the system X = [m b]
52 X = np.linalg.solve(A,B)
53 print('X=',X)
```

**Output:**
```
Location        PovPct  Brth15to17    Brth18to19    ViolCrime    TeenBrth

A= [[9690.44  669.  ]
 [ 669.     51.  ]]
B= [16163.14  1136.4 ]
X= [1.37334539 4.26729284]

Process finished with exit code 0
```

The first 34 lines of the code are the same as before, loading the file and filling the lists for $x$ and $y$ values. The code to construct the $A$ matrix begins at line 36. Line 39 computes $\sum_{i=1}^{n} x_i^2$ and places the value in the first row and first column of matrix $A$. Line 40 sums the $x$ values ($\sum_{i=1}^{n} x_i$) and assigns it to the first row, second column of $A$. Line 42 places the same sum in the second row, first column of $A$. From algebra we know that $\sum_{i=1}^{n} 1 = n$.

Thus, the value to placed in the second row, second column is the number of points in the data set that is equal to the length of the x array. This is accomplished in line 44. The code to build the $B$ matrix begins at 46, placing $\sum_{i=1}^{n} x_i y_i$ in the first element of $B$ and $\sum_{i=1}^{n} y_i$ in the second component of $B$. Finally, the system is solved in line 52, and the results are stored in $X$. The variable $X$ holds two values: the first is $m = 1.37334539$, the second is $b = 4.26729284$. This is a significant program. It accesses a data file, performs arithmetic to build matrices, and solves a linear system. Imagine doing all of that by hand. Then, imagine that the data changes, and you have to do it again.

Let's plot the scatter plot and the regression line together to make sure things are working correctly. We add the following code.

**Code:**

```
54   m = X[0]
55   b = X[1]
56   yhat = m*x+b
57   plt.plot(x,y,'.')
58   plt.plot(x,yhat)
59   plt.legend(['data','regression line'])
60   plt.show()
```

This gives the plot below.

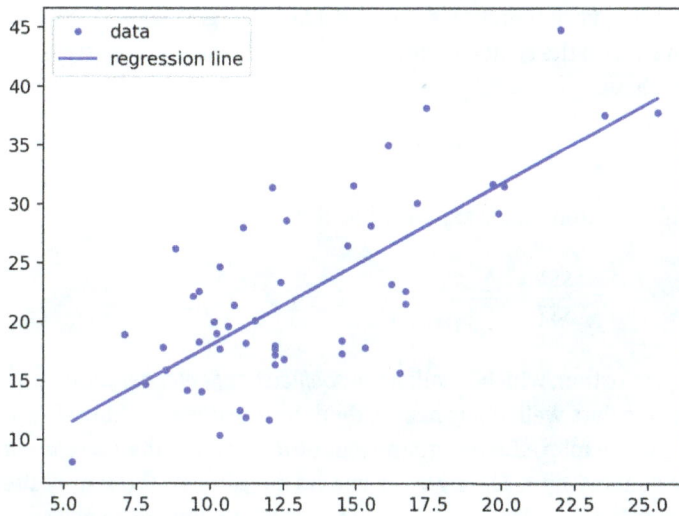

Our regression model is now given by $\hat{y} = 1.373x + 4.267$, and we can use it to predict the average number of births expected for a given poverty percentage. For example, if a certain area had a poverty percentage of 16.2 %, then we would expect

$$\hat{y} = 1.373(16.2) + 4.267 = 26.5096$$

births per 1,000 mothers who are 15–17-years-old in that area.

This seems to be a very nice method to construct linear models. However, there are a couple of questions that still need to be addressed. While we know we have minimized the sum of the squared residuals to derive this line, we do not know whether we should be using a line rather than some other functional form. Also, if we wish to use more than one explanatory variable to aid in the prediction, how do we extend the method to do so?

## 7.1.1 Correlation

To address the question of whether the model should be linear, we use a measure called the *correlation coefficient*. Without working through the derivation, we will rely on the following property:

$$\sum_{i=1}^{n}(y_i - \bar{y})^2 = \sum_{i=1}^{n}(\hat{y}_i - \bar{y})^2 + \sum_{i=1}^{n}(y_i - \hat{y}_i)^2.$$

The sum on the left represents the total variability of the observed values from their mean. This is the same sum that we used when discussing the variation and standard deviation in Chapter 6. We call this the *sum of squares total, SST*. The first sum on the right represents the amount of variation that is explained by the regression. We denote this with *SSR*. The second sum on the right computes the total variation of the errors (or residuals), *SSE*. Hence, we have

$$SST = SSR + SSE.$$

So, the proportion of total variation that is explained by the regression is given by

$$\frac{SSR}{SST} = \frac{\sum_{i=1}^{n}(\hat{y}_i - \bar{y})^2}{\sum_{i=1}^{n}(y_i - \bar{y})^2}.$$

We use $R^2$ to denote this proportion, which is called the *coefficient of determination*. If $R^2$ is near 1, then the line fits the data well. If it is near 0, then the line does not fit well, and other models may need to be explored. The square root of $R^2$ is called the *correlation coefficient* and is usually denoted by $r$. The sign of $r$ will be the same as the sign of the slope of the regression line. The correlation coefficient has the following properties:

- $-1 \le r \le 1$.
- If $r = \pm 1$, then the data are perfectly linear. That is, the data points all lie on the same line. This is nearly impossible in real-world scenarios so be suspicious if such a situation presents itself.

- If $r$ is close to 1 or –1, then the regression line is a good fit for the data.
- If $r$ is near 0, then the data show no linear relationship.

To apply this to our birth rate problem, we simply add the following code.

**Code:**

```
61   SSR = np.sum((yhat-ybar)*(yhat-ybar))
62   SST = np.sum((y-ybar)*(y-ybar))
63   print('SSR=',SSR)
64   print('SST=',SST)
65   R2 = SSR/SST
66   print('R squared=',R2)
67   print('r=',np.sqrt(R2))
```

**Output:**

```
SSR= 1725.2594895969914
SST= 3234.8941176470585
R squared= 0.5333279627871965
r= 0.7302930663693833

Process finished with exit code 0
```

Since the slope of the regression line is positive ($m \approx 1.373$), $r$ is positive. We find that $r \approx 0.73$. This shows that there is definitely a linear relationship between the number of births and the poverty percentage. However, the relationship is not overwhelmingly strong. Whether this value is "strong enough" depends on the context of the problem and the number of data points. In general, the accuracy of the model (prediction) improves as $r$ gets closer to ±1. Finally, $R^2 = 0.5333$ tells us that about 53 % of the variation in the data is explained by the regression line. Again, there is a great deal of theory that we have neglected. If one wants to learn more about linear regression, there are a wealth of probability and statistics textbooks to be consulted that discuss the topic in much more detail than presented here.

**See Exercise 1.**

### 7.1.2 Multiple linear regression

In our example, we currently have only one variable, poverty percentage, that is being used to predict the birthrate. Suppose we wish to also use the crime rate associated with the state to aid in predicting the birth rate. Then, our model would look like this:

$$\hat{y} = m_1x_1 + m_2x_2 + b,$$

where $x_1$ is the poverty percentage, $x_2$ is the crime rate, and $\hat{y}$ is the predicted birth rate. When more than one independent variable is included in the model, we call the process *multiple linear regression*. We could proceed as we did previously and would now need to take partial derivatives with respect to $m_1, m_2$, and $b$. Set each of the partial derivatives equal to zero and solve the resulting system. That is an acceptable approach. But, what if we wanted to use five predictor variables, or ten? You can see how this might become cumbersome. It may be that, as we develop those systems, we would see a pattern evolve that would speed our work. In fact, that is likely the case. But there is another approach that achieves the same regression line by using a matrix representation of the regression process. To reduce notational complexity, let's let $p$ be the poverty rate and $c$ the crime rate. Then, if our data were perfectly linear, we would have that for each point:

$$y_i = m_1p_i + m_2c_i + b.$$

For convenience, we will write the intercept first to have

$$y_i = b + m_1p_i + m_2c_i.$$

Now, we construct matrices to hold the $y$ values and the predictor variables. So, we let

$$\mathbf{Y} = \begin{bmatrix} y_1 \\ y_2 \\ \vdots \\ y_n \end{bmatrix}, \quad \mathbf{A} = \begin{bmatrix} 1 & p_1 & c_1 \\ 1 & p_2 & c_2 \\ & \vdots & \\ 1 & p_n & c_n \end{bmatrix}, \quad \text{and} \quad \mathbf{S} = \begin{bmatrix} b \\ m_1 \\ m_2 \end{bmatrix}.$$

Then, the set of equations for perfect data would be expressed as $\mathbf{AS} = \mathbf{Y}$. Now, if $\mathbf{A}$ were a square matrix, we could solve this as we have earlier systems. However, $\mathbf{A}$ is not square, and, in the actual data, we could have differing $y$ values for the same $(p, c)$ pair. Thus, taking the inverse of $\mathbf{A}$ is not likely to be possible. To address this "non-squareness," we multiply both sides of the equation by the transpose of $\mathbf{A}$ to get

$$\mathbf{A}^T\mathbf{AS} = \mathbf{A}^T\mathbf{Y}.$$

If we were to examine this system, we would find that this generates the same equations that are generated by taking the derivatives and moving constant terms to one side (which seems incredible to me). But, since it is in matrix form, we can solve it using matrix methods as before. So, let $\mathbf{Q} = \mathbf{A}^T\mathbf{A}$. Then we have

$$\mathbf{QS} = \mathbf{A}^T\mathbf{Y}.$$

Finally, we simply multiply both sides by the inverse of $\mathbf{Q}$ to solve for the coefficients in the regression model, i. e.,

$$QS = A^T Y$$
$$Q^{-1}QS = Q^{-1}A^T Y \qquad (7.1)$$
$$S = Q^{-1}A^T Y.$$

We will apply this idea to our current model. We want to use poverty rate and crime rate as the independent (explanatory) variables and birth rate as the dependent (response) variable. Thus, **A** would have a column of ones, a column holding the poverty rates, and a column holding the crime rates. Once **A** has been constructed, we let Python do the rest as shown next.

**Code:**

```
1   import numpy as np
2
3   #open the data file
4   pov_file = open('poverty.txt','r')
5   #see how many lines are in the file
6   count = len(pov_file.readlines())
7   pov_file.close()
8   #since the first line contains headers, there is one less actual
9   #lines of data
10  count = count-1
11
12  #set up some storage space for the data
13  A = np.ones((count,3))
14  y = np.zeros((count,1))
15  #now we will reopen the file and read it line by line
16  #the first line of this file is a header
17  pov_file = open('poverty.txt','r')
18  headers = pov_file.readline()
19  #i printed the headers just in case i wanted to reference them
20  print(headers)
21  #now read the rest of the file
22  for i in range(count):
23      #get the next line
24      l = pov_file.readline()
25      #split the line into separate fields
26      fields = l.split()
27      #the second field is the poverty percent
28      A[i,1] = float(fields[1])
29      #the fifth field is the crime rate
30      A[i,2] = float(fields[4])
31      #the third field holds the births that we want.  store them in y
32      y[i] = float(fields[2])
33  #close the file
```

```
34   pov_file.close()
35   #multiply both sides by A_transpose
36   A_trans = A.transpose()
37   Q = np.dot(A_trans,A)
38   #right hand side
39   RHS = np.dot(A_trans, y)
40
41   #now solve the system
42   X = np.linalg.solve(Q,RHS)
43   print('X=',X)
44   print('yhat = {:.4f} + {:.4f}(poverty) + {:.4f}(crime)'\
45         .format(X[0,0],X[1,0],X[2,0]))
46
47   ybar = np.average(y)
48
49   yhat = np.dot(A,X)
50
51   SSR = np.sum((yhat[:,0]-ybar)*(yhat[:,0]-ybar))
52   SST = np.sum((y-ybar)*(y-ybar))
53   print('SSR=',SSR)
54   print('SST=',SST)
55   R2 = SSR/SST
56   print('R squared=',R2)
57   print('r=',np.sqrt(R2))
```

Notice that in line 13 we initialize the data matrix, **A**. It is established with three columns and preloaded with ones instead of zeros. Then, we can leave the first column as is and adjust the second and third columns as needed. These data values are assigned in lines 28 and 30. The *y* values are as they were before. To accomplish the regression, we use equation (7.1). We take the transpose of the data matrix in line 36 and multiply the transpose by the data matrix in line 37 to give us **Q**. We must also multiply the right-hand side of the equation by $\mathbf{A}^T$, and this is done in line 39, creating **RHS**. Then, we solve as usual in line 42. The vector **X** holds the intercept and coefficients of the model. The result is displayed in the output.

**Output:**

| Location | PovPct | Brth15to17 | Brth18to19 | ViolCrime | TeenBrth |
|---|---|---|---|---|---|

```
X= [[5.98220133]
 [1.03649967]
 [0.34420732]]
yhat = 5.9822 + 1.0365(poverty) + 0.3442(crime)

Process finished with exit code 0
```

We can compute the coefficient of determination in a similar fashion to the previous example. We add the following code.

**Code:**

```
57   ybar = np.average(y)
58   yhat = np.dot(A,X)
59   SSR = np.sum((yhat[:,0]-ybar)*(yhat[:,0]-ybar))
60   SST = np.sum((y-ybar)*(y-ybar))
61   print('SSR =',SSR)
62   print('SST =',SST)
63   R2 = SSR/SST
64   print('R squared =',R2)
```

**Output:**

```
SSR = 2092.1951714030574
SST = 3234.8941176470585
R squared = 0.6467584703900114

Process finished with exit code 0
```

Using this matrix approach, we can include as many predictor variables as we wish by adding columns to the data matrix. The method is unchanged regardless of the number of variables.

One of the advantages to using a language like Python is that other smart people also use the language. That being the case, libraries and packages are developed and made available all the time. Thus, there are available packages that will compute the regression line for us. There are many such packages, but I have chosen to demonstrate the third-party library found in *scikit*. So if you do not have scikit installed, you will need to add it to your Python installation as we have done with other packages. The *scikit* package includes sklearn, which contains the regression methods we wish to use. We will use this package to perform the multiple regression that we previously accomplished, using poverty rate and crime rate as the independent variables. In addition, we can use the genfromtxt method that we introduced in Section 6.2 to speed the process of loading the data into matrices. The code to construct the regression is next.

**Code:**

```
1   import numpy as np
2   from sklearn.linear_model import LinearRegression
3
4   # load the poverty percent into column 1 of A and the crime rate in column 2
5   # we do not need a column of ones because the method will do that for us
6   A = np.genfromtxt('poverty.txt',dtype=float,usecols=(1,4), skip_header=True)
7   # load the birth rates into Y
```

```
8   Y = np.genfromtxt('poverty.txt',dtype=float,usecols=(2), skip_header=True)
9
10  # now the data matrix A and the actual y values Y are complete
11  # fit the regression line and store it.
12  # we are storing it in a variable named birthmodel of type LinearRegression
13  birthmodel = LinearRegression()
14  # find the parameters for the regression line
15  birthmodel.fit(A, Y)
16  # get the coefficient of determination (R-squared_
17  R2 = birthmodel.score(A, Y)
18  print('R squared =',R2)
19  # variables of type LinearRegression have components called
20  # coef_ and intercept_ that store the coefficients and intercept of
21  # the model.
22  coeff = birthmodel.coef_
23  intercept = birthmodel.intercept_
24  print('yhat = ({:.4f})poverty + ({:.4f})crime + {:.4f}'\
25          .format(coeff[0],coeff[1],intercept))
```

If one examines this code, we see that there are only 11 lines of actual executable statements. The rest are comments. So, we built an entire multiple regression code with 11 statements. Hopefully, we are beginning to see the power of a programming language. Let's look at the code a bit.

In line 2, we use a different form of the import statement. The *scikit* package is very large, and it would be inefficient to import the entire package if we only need a small part of it. Thus, we take from it only the linear regression module that we need, which is called *LinearRegression*. Line 6 loads the poverty and crime rates in the matrix A. We can see that columns 2 and 5 (indices 1 and 4) of the data file are used and that the header line is skipped. Since *poverty.txt* is space or tab delimited, we do not need to specify a delimiter in genfromtxt as we did in Chapter 6. In line 8, the birth rates for 15–17-year-olds are loaded into a vector Y. When we programmed our own regression method, we had to include a column of ones in the data matrix. That is not necessary here since the methods in *scikit* will do that for us when needed. Line 13 sets up a variable to hold the results of the linear regression. The type of variable here is LinearRegression. This is the same as saying that we are setting aside space for a package of items, and the package is called birthmodel. The package has a prebuilt method that will do all the same work that we have done in our previous regression example. The method is called .fit, and we build the regression model in line 15 with birthmodel.fit(A,Y), indicating the data matrix and the actual *y* values associated with the data. Once this line has been executed, the regression is complete, but we have to know how to access the information. Such access comes via other items in the package. To find the coefficient of determination, we use the .score item, as in line 17. Remember, our regression model will look like

$$\hat{y} = (c_1)\text{poverty} + (c_2)\text{crime} + b,$$

where $c_1$ and $c_2$ are the coefficients of the variables and $b$ is the intercept of the model. To access the coefficient values, we use the `coef_` method. It will return a vector of coefficients that were computed when `.fit` was executed. Line 22 assigns this vector of coefficients to the variable `coeff`. Similarly, the intercept ($b$) is found by using the `.intercept_` method and assigned to the variable `intercept` in line 23. The `coeff` and `intercept` variables are used to print the regression model that was found in line 26. The output of the program is next.

**Output:**
```
R squared = 0.6467584703900118
yhat = (1.0365)poverty + (0.3442)crime + 5.9822


Process finished with exit code 0
```

We can see that the regression model and the $R^2$ value are the same as we found before. If we wanted to compute $\hat{y}$ for particular values of poverty rate and crime rate, we could use the `.predict` method. Suppose that $p = 15$ and $c = 10$. Then, we could compute the prediction with

```
yhat = birthmodel.predict([[15,10]])
```

This is equivalent to $1.0365(15)+0.3442(10)+5.9822$. Note that the argument of the method is a matrix. That is why there are two pairs of square brackets. If we wanted the predicted values for all of the data points, we could use the following command:

```
yhat = birthmodel.predict(A)
```

When we have more than one explanatory variable, we can no longer use scatterplots to "visualize" the data because we leave two dimensional space (i. e., we have more than just $x$ and $y$). Thus, we rely more on measures like $R^2$ to tell us whether the linear model is appropriate. Also, once we have a linear model, the absolute value of the coefficients indicates which variable has more impact on the prediction. In our example, because 1.0365 is substantially greater than 0.3442, we can say that poverty rate plays a larger role than crime rate in the prediction of birth rates.

There are many other methods available that enable one to study the residuals and other measures of fitness. Using a third-party package like *scikit* often requires the development of much less code. Instead, the effort is focused on installing and implementing the package and on accessing and using the model that is produced. The motivated student will find a wealth of available theory and application if linear regression is to be pursued further.

**See Exercises 2 and 3.**

## 7.2 Logistic regression

When conducting linear regression, we have a continuous dependent (target) variable, i. e., the variable is allowed to take on any value within the data range. For example, in our birth-rate example, the birth rate could be any positive percentage. We were not restricted to whole number percentages or had any other constraint. However, there are many situations in which such freedom is not present. Suppose we poll 50 athletes to obtain their heights and the sports they play. Further, we speculate that "tall" athletes are more likely to play basketball. Then, our response variable is an indicator of whether the athlete plays basketball. Thus, there are only two choices for the variable: yes (they play basketball) or no (they do not). We review the data and find that 25 of them play basketball. The results of the poll are in the following table.

| Heights (in) of Basketball Players | Heights (in) of Non-Basketball Players |
|---|---|
| 78.29 | 74.22 |
| 82.28 | 71.2 |
| 81.36 | 73.85 |
| 80.27 | 74.2 |
| 76.48 | 74.35 |
| 78.33 | 68.99 |
| 82.05 | 69.45 |
| 80.71 | 70.77 |
| 76.02 | 76.87 |
| 79.44 | 73.29 |
| 70.87 | 70.61 |
| 75.32 | 75.85 |
| 79.46 | 73.8 |
| 74.52 | 71.01 |
| 77.8 | 72.65 |
| 74.9 | 76.8 |
| 78.97 | 74.39 |
| 81.12 | 67.63 |
| 83.04 | 70.95 |
| 78.59 | 69.17 |
| 75.73 | 69.94 |
| 77.59 | 71.79 |
| 75.69 | 72.04 |
| 78.72 | 70.69 |
| 76.32 | 68.4 |

We can use a value of 1 to indicate that an athlete plays basketball and value of 0 to indicate that an athlete does not play basketball. Let's use the variable $y$ to indicate the playing status. So, we see that 0 and 1 are the only two possible values for $y$. There is no such thing as $y = 0.5$, or $y = 0.23$, or any other value. This means that $y$ is not continuous but *discrete*. When a variable is discrete, we can list the possible values of the variable, and there is space between the numeric values. For our current example, we can plot the data using the height as the first coordinate and the basketball status $y$ as the second coordinate. We get the following scatterplot.

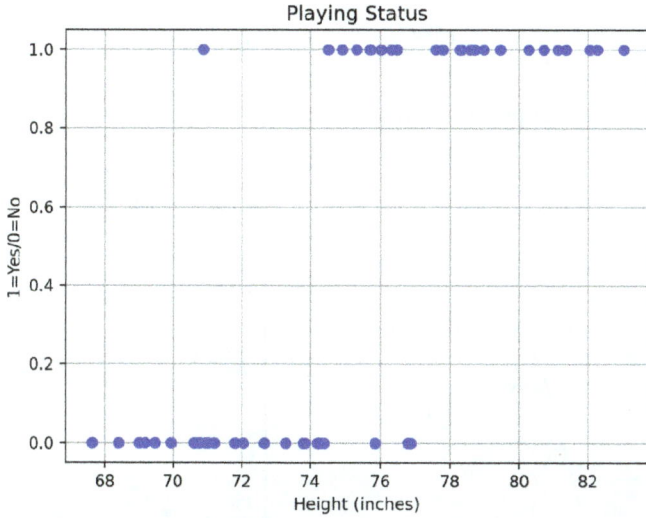

We could proceed to conduct linear regression to find the line of best fit as shown in the next graph.

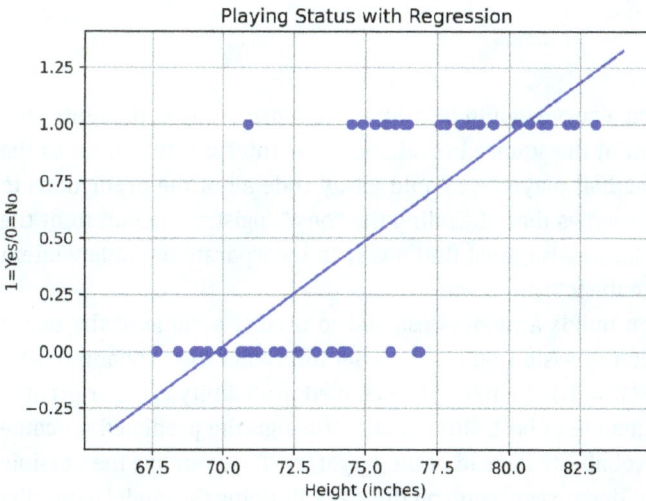

This regression line yields a coefficient of determination of $R^2 = 0.5604$. While this seems to be a moderately successful regression, it is clear that these data are not linear. Furthermore, since the target should be 0 or 1, the prediction for most values will not be very close to either of those values. The regression could also yield predictions that are much above 1 or below 0, which are, in theory, meaningless for this scenario. Instead, we might seek to use a model that is not linear and that better attempts to be 0 or 1. One such curve is the logistic function,

$$f(x) = \frac{e^{g(x)}}{1 + e^{g(x)}}.$$

An example of a logistic function is superimposed on the previous regression results in the following graph.

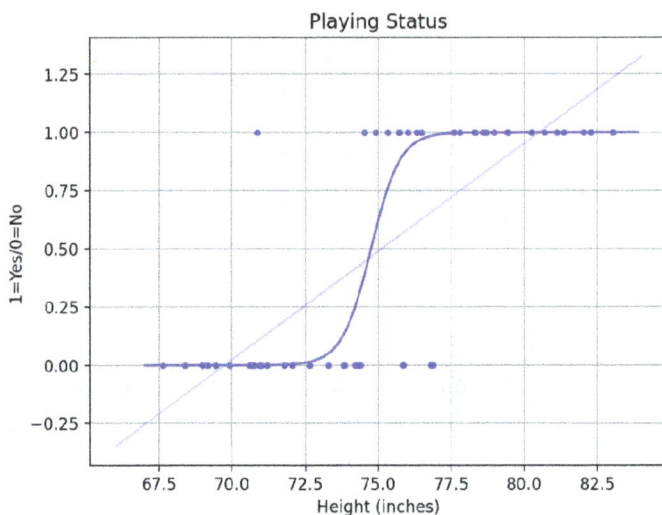

One can easily see that such a curve would be "closer" to more of the data points and, thus, would reduce the sum of the squared residuals. If we interpret the curve as the probability of being a basketball player, we could easily code all of the predictions to be either 0 or 1. The task becomes that of finding the "best" logistic function to fit the data. Within that broad task, we also must find a way to incorporate the independent (explanatory) variables into the logistic function.

While linear regression builds a model designed to predict a value of the target (response) variable, logistic regression builds a model to predict the probability of a particular outcome (e. g., $P(Y = 1)$). Then, if the predicted probability is 0.5 or greater, we assign the predicted outcome to be 1. Otherwise, we assign the predicted outcome to be 0. By predicting the probability instead of the target itself, we remove the possible bias that could be present in linear regression on this data. Building the model to predict the probability is much more complicated than in the case of linear regression.

While the details of the process are beyond the scope of this text, we supply some background for those students who wish (or need) to know some of the theory in order to understand the broader picture. Using our basketball example, let the height be denoted by $x$ and the basketball status by $y$. We let 1 indicate that the athlete is a basketball player. Then, we denote the probability of being a basketball player by $p$. Thus, $P(y = 1) = p$. The odds of any event $E$ are given by

$$\text{Odds}(E) = \frac{\text{probability of } E}{1 - (\text{probability of } E)} = \frac{P(E)}{1 - P(E)}$$

So, for our example, we have

$$\text{Odds}(\text{playing basketball}) = \frac{P(y = 1)}{1 - P(y = 1)}$$
$$= \frac{p}{1 - p}.$$

The process of logistic regression attempts to find a curve that best fits the Odds *of* $E$ as the target. However, it is clear that the odds of an event is not linear. Since the odds formula involves a fraction (which includes $p$), we can decompose the fraction by taking the natural log of the odds. It can be shown that we can model $\ln(\text{odds } of \ E)$ in a linear fashion. In the context of our example, we can find coefficients so that

$$\text{Predicted}(\ln(\text{Odds of } E)) = c_0 + c_1(\text{height}).$$

This may seem a bit arbitrary, but consider the following. Assume we fit the model perfectly so that

$$\ln(\text{Odds of } E) = c_0 + c_1(\text{height}).$$

Then, if $x = $ height, we have

$$\ln\left(\frac{p}{1 - p}\right) = c_0 + c_1 x$$
$$e^{\ln\left(\frac{p}{1-p}\right)} = e^{c_0 + c_1 x}$$
$$\frac{p}{1 - p} = e^{c_0 + c_1 x}$$
$$p = (1 - p)e^{c_0 + c_1 x}$$
$$p = e^{c_0 + c_1 x} - pe^{c_0 + c_1 x}$$
$$p + pe^{c_0 + c_1 x} = e^{c_0 + c_1 x}$$
$$p(1 + e^{c_0 + c_1 x}) = e^{c_0 + c_1 x}$$
$$p = \frac{e^{c_0 + c_1 x}}{1 + e^{c_0 + c_1 x}}.$$

The last equation shows that, if the log of the odds is linear, then the model for $p$ is a logistic function (and we know the parameters of said logistic function). Very nice! Thus, we seek to find the "best" values for $c_0$ and $c_1$ to define the predicted value of $p$,

$$\hat{p} = \frac{e^{c_0 + c_1 x}}{1 + e^{c_0 + c_1 x}}.$$

So, can we do linear regression from this point on? We might be tempted to do so, using the log of the odds as the target variable. The problem is that we need values for the odds of each of our observations. For example, one of our (height, basketball status) pairs is (78.29, 1). So we have a player who is 78.29 inches tall and plays basketball. Since we know he plays basketball, we might say that the probability of his playing basketball is 1. If we do so, the odds of his playing basketball would be $\frac{1}{1-1} = \frac{1}{0} = \infty$. Obviously, this poses a problem when trying to compute regression coefficients. Because of this, logistic regression does not attempt to minimize the sum of the squared residuals. Instead, it tries to maximize the probability that all of the observations have occurred. This maximization requires that we construct what is called a joint probability function. To do so, you need to know a bit of probability that is not covered in this text, but we would construct a new function called the *likelihood function* that has $c_0$ and $c_1$ as variables. The likelihood function would also incorporate all of the observations from the data. Then, as we did with linear regression, we would take derivatives with respect to each variable. Finally, we would need to solve the resulting system of equations. It sounds straightforward. However, the resulting system of equations is such that we cannot use the methods discussed in this test. We would have to use a much more complicated solution technique. It is hoped that we are making it clear that logistic regression is a much more complex process than linear regression. So, we should definitely look for a package or library to accomplish this.

The sklearn package (from scikit) contains methods that will perform logistic regression, similar to the linear regression methods we used in the previous section. The code to perform the logistic regression for our basketball example is given next.

**Code:**

```
1   import numpy as np
2   import matplotlib.pyplot as plt
3
4   #load the required tools for logistic regression
5   from sklearn.linear_model import LogisticRegression
6   from sklearn import metrics
7
8   #create space for the height values.  This must be a matrix, even
9   #if there is only one variable.
10  x = np.zeros((50,1))
11  x[:,0] = np.array([78.29, 82.28, 81.36, 80.27, 76.48, 78.33, 82.05, 80.71,
12    76.02, 79.44, 70.87, 75.32, 79.46, 74.52,  77.80, 74.90, 78.97, 81.12,
13    83.04, 78.59,  75.73, 77.59, 75.69,  78.72, 76.32,74.22, 71.20, 73.85,
14    74.20, 74.35, 68.99, 69.45, 70.77, 76.87, 73.29, 70.61, 75.85, 73.80,
15    71.01, 72.65, 76.80, 74.39,  67.63, 70.95,  69.17, 69.94, 71.79, 72.04,
16    70.69, 68.40])
17
18  #create storage space for status values
19  #this initializes with 1's in all the values
```

```
20  y = np.ones(50)
21  #the second 25 values in the data are not basketball players
22  #so the status values should be replaced with 0's
23  y[25:50] = np.zeros(25)
24
25  #establish a regression object
26  BasketballRegr = LogisticRegression()
27
28  #perform the regression to fit the model
29  BasketballRegr.fit(x, y)
30
31  #find the percentage of correct classifications
32  score = BasketballRegr.score(x, y)
33  print('Percentage of correct predictions:',score*100)
34
35  #calculate the predicted values
36  yhat = BasketballRegr.predict(x)
37
38  #print the confusion matrix
39  print('Confusion Matrix')
40  cm = metrics.confusion_matrix(y, yhat)
41  print(cm)
```

The code is documented (commented) to explain what each new command is to accomplish. Lines 5 and 6 import the tools necessary to do the regression and to make accessible certain metrics that measure the goodness of the results. The regression allows for any number of independent (explanatory) variables. Thus, the values of the variables are to be stored in a matrix. While our basketball example has only one independent variable, height, we still must put it into a matrix, instead of a vector. The matrix is initialized with zeros in line 10. In line 11, the 50 heights are placed into the matrix called x. The first 25 values represent the heights of the basketball players, while the remaining 25 are for those who did not play basketball. We build the vector of status values in lines 20 and 23. The y vector is initialized with ones. Then, the last 25 entries are replaced by zeros. To perform the regression, we need a regression object much like we used in the previous section. Line 26 initializes a variable called BasketballRegr to hold the results of the regression. Line 29 actually performs the regression to find the optimal values for the coefficients in the logistic model. At this point the regression is done, but we need to know how to use and access the results. Line 32 accesses a method called .score in the regression object that will report the number of predictions that are correct, given the x and y values. In our case, the model predicts the player status correctly 88 % of the time, missing the prediction on 6 of the 50 values. To get more detail on the predictions, we can create a *confusion matrix*. The confusion matrix indicates how many of each category were predicted correctly. So, we need to get the predictions for our data and compare them to the actual values. The predictions are obtained via line 36 with the .predict method. The argument of .predict is either a single $x$ value or a matrix of $x$ values. The result gives the predictions for each of the $x$ values. The confusion matrix for this regression is given here.

| | | Predicted Value | |
|---|---|---|---|
| Actual Value | | 0 | 1 |
| 0 | | 22 | 3 |
| 1 | | 3 | 22 |

The row numbers of the matrix represent the actual values of the target variable, and the column numbers represent the predicted value. Let's call the confusion matrix $C$. Then, for our example $C(0, 0) = 22$ means that there were 22 observations for which the actual target value and the predicted target value were both 0. Hence, these 22 predictions were correct. In the first column, $C(0, 1) = 3$, we see that there were three values for which the actual value was 0, but the predicted value was 1. Thus, these three predictions were incorrect. Likewise, in the next row, we have $C(1, 0) = 3$ and $C(1, 1) = 22$. This indicates the there were 22 correct predictions when the target value was 1 and three incorrect predictions. The values along the main diagonal of $C$ are those predictions that were correct, while any value off of the diagonal represents an incorrect prediction. The confusion matrix is not necessarily symmetric. It just happens to be in this case.

To get the actual confusion matrix, we use the metrics.confusion_matrix(y, yhat) command. The command assigns the results to a matrix in line 40. We called the matrix cm, but it can take any valid variable name. The output from the previous code is given next.

**Output:**

```
Percentage of correct predictions: 88.0
Confusion Matrix
[[22  3]
 [ 3 22]]


Process finished with exit code 0
```

If we wish to make a prediction for a new observation, we can again use the .predict method. Suppose we wish to predict whether an athlete with a height of 72 inches plays basketball. Then, we add the following code.

```
newy = BasketballRegr.predict([[72]])
print('Status for this player: ',newy)
```

We need [[72]] because the method expects a matrix. The output follows:

```
Status for this player: [0.]
```

Thus, our model predicts that an athlete who is 72-in tall would be assigned to status 0, which indicates that the athlete does not play basketball.

Finally, if we wish to know the coefficients that are actually found by the regression we use two other methods: .coef_ and .intercept_. We could use the following code:

```
coeff = BasketballRegr.coef_
c_1 = coeff[0]
c_0 = BasketballRegr.intercept_
print('c_0 = ',c_0)
print('c_1 = ',c_1)
```

**Output:**
```
c_0 = [-56.92020903]
c_1 = [0.75782544]
```

Since we could have more than one independent variable, the .coef_ method returns a vector of values. From the output we can determine the model for log of the odds to be

$$\ln\left(\frac{p}{1-p}\right) \approx -56.92020903 + 0.75782544x,$$

where $x$ is height. We can solve this for $p$ to get the logistic model

$$\hat{p} = \frac{e^{-56.92020903+0.75782544x}}{1 + e^{-56.92020903+0.75782544x}}.$$

Remember that, if $\hat{p} \geq 0.5$, our prediction is 1. Otherwise, the prediction is 0. The corresponding logistic curve is shown in Figure 7.2.

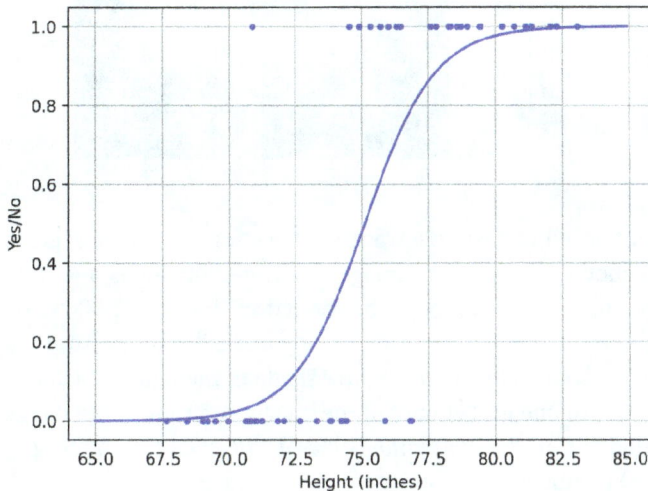

**Figure 7.2:** Original scaterplot with logistic model.

The scatterplot represents a simple example of logistic regression. In practice, we use logistic regression with much larger data sets. Also, logistic regression allows for more than two classes. In the next section, we expand upon our discussion by considering a much larger, more complicated scenario.

### 7.2.1 Digit recognition model

While the previous example involved one variable, two classes (basketball or not), and 50 observations, we generally use logistic regression on much larger sets of data involving many variables and multiple classes. Logistic regression extends naturally, though tediously, to allow for almost any number of classes and variables. Fortunately, most of the complexity of this extension is hidden in the logistic-regression methods that are provided. To demonstrate the power of logistic regression, we consider a sample data set that is included in the sklearn package. The data set is called *digits* and it includes a large set of images of the numbers 0, 1, 2, 3, 4, 5, 6, 7, 8, and 9. The image data is represented by an 8×8 matrix in which each element of the matrix represents an intensity (or shade) of that part of the image. For example, one of the matrices and its corresponding image are given next.

$$D = \begin{bmatrix} 0 & 0 & 0 & 2 & 13 & 0 & 0 & 0 \\ 0 & 0 & 0 & 8 & 15 & 0 & 0 & 0 \\ 0 & 0 & 5 & 16 & 5 & 2 & 0 & 0 \\ 0 & 0 & 15 & 12 & 1 & 16 & 4 & 0 \\ 0 & 4 & 16 & 2 & 9 & 16 & 8 & 0 \\ 0 & 0 & 10 & 14 & 16 & 16 & 4 & 0 \\ 0 & 0 & 0 & 0 & 13 & 8 & 0 & 0 \\ 0 & 0 & 0 & 0 & 13 & 6 & 0 & 0 \end{bmatrix}$$

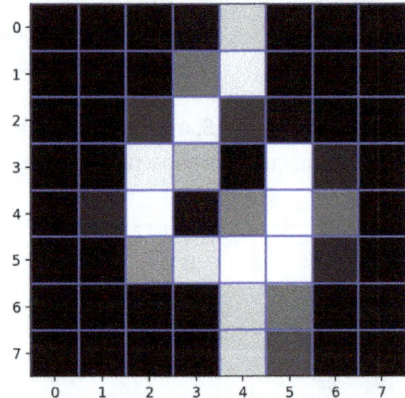

The blue grid is artificially imposed to ease the process of locating particular regions of the image. The matrix $D$ above represents an image that is designated as a 4. Note that $D(0,0) = 0$ corresponds to a black square at the top left of the image. Likewise, $D(3,2)$ corresponds to the square in the $4^{th}$ row from the top and $3^{rd}$ column from the left. The value is 15, which corresponds to a much lighter shade in the image. Each element of the matrix corresponds to one grid square of the image. To produce the image in Python takes a new method within the *matplotlib* package. The code to produce the image (without the grid lines) is shown here.

**Code:**

```
1   import numpy as np
2   import matplotlib.pyplot as plt
3   #plot a sample image
4   D = np.matrix([
5    [ 0.,  0.,  0.,  2., 13.,  0.,  0.,  0.],
6    [ 0.,  0.,  0.,  8., 15.,  0.,  0.,  0.],
7    [ 0.,  0.,  5., 16.,  5.,  2.,  0.,  0.],
8    [ 0.,  0., 15., 12.,  1., 16.,  4.,  0.],
9    [ 0.,  4., 16.,  2.,  9., 16.,  8.,  0.],
10   [ 0.,  0., 10., 14., 16., 16.,  4.,  0.],
11   [ 0.,  0.,  0.,  0., 13.,  8.,  0.,  0.],
12   [ 0.,  0.,  0.,  0., 13.,  6.,  0.,  0.]])
13
14  plt.imshow(D,cmap = plt.cm.gray)
15  plt.show()
```

The values of the matrix are loaded into a matrix named D. Then, we use the .imshow method to construct the image. There is a parameter called cmap which designates the color scheme to use when shading the image. There are many maps to choose from, or the user can define one. A couple of other maps are cmap = plt.cm.summer and cmap = plt.cm.Purples. The documentation for matplotlib provides information about many more available color maps.

The *sklearn* package provides a large database of images and the digits to which they are assigned. We want to use logistic regression to build a model that would predict the digit, based on a provided image. To load the image data, we use the following.

**Code:**

```
1   from sklearn.datasets import load_digits
2
3   #get the data for all the images
4   digits = load_digits()
5
6   # Determine the size of the image data
7   print('Image Data Shape' , digits.data.shape)
8   # Make sure the targets are appropriate size
9   print('Label Data Shape', digits.target.shape)
```

**Output:**

```
Image Data Shape (1797, 64)
Label Data Shape (1797,)


Process finished with exit code 0
```

Now, this method of loading the data is unique to this situation. As we have seen, data may be found in many places and in many formats. We have to deal with finding and processing data as the situation demands. In this case, the data is loaded by a method in the sklearn package. We assign the data to a name (we chose digits) via the load_digits() command on line 4. We then have a variable called digits that is a new type of variable (whatever load_digits gave us). The digits variable has some methods available that we use to find the shape of the digits data and the associated assigned digits that go with each data record. From the output we see that the image data is stored in a matrix that is 1797 by 64. So, each row of digits contains the intensity values of an 8 × 8 matrix. We also see that there are 1,797 images represented. In sklearn, the y values (response values) are called targets. With the .target.shape method, we see that we have 1,797 corresponding labels that tell us which digit each image it is supposed to represent.

In this scenario, we will treat each of the 64 positions in the matrix as a variable (64 different locations in the image). We have ten categories, one for each digit 0 through 9, and we have 1,797 observations. We wish to perform logistic regression to build a model from this data. With a large set of data, it is common to split the data into two sets. We use one set to actually do the regression. We use the other set to act as new data to see how well the model predicts for data that it has not seen before. The data that is used to perform the regression is called the *training set*, while the data that is used to test the accuracy of the model is called the *testing set*. We could do this split ourselves, but sklearn has provided routines that do this for us. Code to split the digits data is given next.

**Code:**
```
10   from sklearn.model_selection import train_test_split
11   x_train, x_test, y_train, y_test = train_test_split(digits.data,
12        digits.target, test_size=0.25, random_state=1)
```

The first of these commands imports the needed method from the sklearn package. The second splits the data and the targets into two sets. The explanatory values are loaded into matrices named x_train and x_test, and the response values (targets) are loaded into vectors named y_train and y_test. The x denotes the independent variables, and the y denotes the target variables. Note that we are not required to use these names. Any valid variable name is allowed, but, as usual, best practices dictate that the variable name be meaningful if possible. By setting test_size=0.25, we use 75 % of the data to "train" or fit the model and the remaining 25 % will be used to test the model to see how well it performs. The records in the data are chosen randomly for the split, and the random_state parameter enables us to control the selection to some degree. If random_state is set to be an integer, then the same split of training and testing data will be done each time the program is run so that the results will be reproducible. Different integers will select different random sets. If the parameter is not included, then a different random set of data will be selected each time the program is run. Once the data has been split, we should be able to run the logistic regression on the training data. We do this similarly to the way we did the basketball problem. The proposed code is given next.

**Code:**
```
1    from sklearn.model_selection import train_test_split
2    from sklearn.datasets import load_digits
3    from sklearn.linear_model import LogisticRegression
4
5    #get the data for all the images
6    digits = load_digits()
7
8    # Determine the size of the image data
9    print('Image Data Shape' , digits.data.shape)
10   # Make sure the targets are appropriate size
11   print('Label Data Shape', digits.target.shape)
12
13   x_train, x_test, y_train, y_test = train_test_split(digits.data, \
14                        digits.target, test_size=0.25, random_state=1)
15
16   # all parameters not specified are set to their defaults
17   logisticRegr = LogisticRegression()
18
19   #do the logistic regression
20   logisticRegr.fit(x_train, y_train)
21   # Use score method to get accuracy of model
22   score = logisticRegr.score(x_test, y_test)
23   print('Percent correct =', score)
```

Lines 17, 20, and 22 are the same methods we used in our basketball example to perform the logistic regression, except that the .score method is applied to the test data instead of the data that was used to build the regression mode. The code attempts to fit the regression on the training data and print the percentage of the test data that is predicted correctly. However, the following output is given.

**Output:**
```
Image Data Shape (1797, 64)
Label Data Shape (1797,)
/Library/Frameworks/Python.framework/Versions/3.8/lib/python3.8/
 site-packages/sklearn/linear_model/_logistic.py:938:
 ConvergenceWarning: lbfgs failed to converge (status=1):
STOP: TOTAL NO. of ITERATIONS REACHED LIMIT.
```

```
Increase the number of iterations (max_iter) or scale the data as shown
    in: https://scikit-learn.org/stable/modules/preprocessing.html
Please also refer to the documentation for alternative solver options:
    https://scikit-learn.org/stable/modules/linear_model.html#logistic-
regression
  n_iter_i = _check_optimize_result(
Percent correct = 0.9688888888888889

Process finished with exit code 0
```

The last line indicates that the process finished with exit code 0, and the score of 0.968888 was displayed, indicating that an effective regression model was constructed. However, we received a STOP message. Obviously, this is something we should not ignore. Reading through the message, it says that the maximum number of iterations was reached. Further, it says that we could increase the parameter `max_iter` or scale the data. We will look into each of these options. As we discussed earlier, optimizing the likelihood function for logistic regression cannot be done by solving a linear system, as we did previously in this book. Instead, the solution is achieved via an iterative process. The process starts with an initial guess. Then, each iteration aims to improve upon the previous solution (much like we saw in the bisection method presented earlier in the text). The solution after each iteration is called an *iterate*. The process will continue as long as the iterates are getting better by a significant amount. So, in the previous case, there is a limit on the number of iterations, and the process had not reached what it believed to be the best solution. Thus, we need to do some research to see how to change this maximum number of iterations. Doing such investigation, we find that the `max_iter` parameter is established in the initialization of the logistic regression (line 17). We can modify that initialization as follows.

**Code:**

```
17  logisticRegr = LogisticRegression(max_iter=2800)
```

Using trial and error, we can find the number of iterations that allows the process to conclude without the warning.

**Output:**

```
Image Data Shape (1797, 64)
Label Data Shape (1797,)
Percent correct = 0.9688888888888889

Process finished with exit code 0
```

The next question, from a computing perspective, is whether we can reduce the number of iterations needed. Fewer iterations would mean that the process takes less time. Without much knowledge as to how the solver is working, we look to the other part of the warning that was previously given. We could try to scale the data. Scaling numeric data is common. If numbers within a data set differ by several orders of magnitude, it often leads to numerical issues when trying to compute with such numbers. Often rounding and truncation error become a problem, and iterative solvers could also struggle. There are many ways to scale data, but one of the easiest and most common ways is to simply divide all values by the same number. Frequently, we find the largest absolute value within the data and divide all numbers by that maximum. Then, the scaled values are all between –1 and 1. If the values are all positive, then the scaled values are all between 0 and 1. We could do such scaling in Python, but sklearn has provided methods to do many different types of scaling. So, we will try the default scaling that is offered. Once again, we need another package from sklearn called *preprocessing*. To scale the data for our current example, we modify the previous code as follows.

**Code:**

```
1   from sklearn.model_selection import train_test_split
2   from sklearn.datasets import load_digits
3   from sklearn.linear_model import LogisticRegression
4   from sklearn import preprocessing
5
6   #get the data for all the images
7   digits = load_digits()
8
9   #scales the data to help with numeric computation
10  data_scaled = preprocessing.scale(digits.data)
11
12  # Print to show there are 1797 images (8 by 8 images for a dimensionality of 64)
13  print('Image Data Shape' , digits.data.shape)
14  # Print to show there are 1797 labels (integers from 0-9)
15  print('Label Data Shape', digits.target.shape)
16
17  x_train, x_test, y_train, y_test = train_test_split(data_scaled, digits.target,\
18                                          test_size=0.25, random_state=0)
19
20  # all parameters not specified are set to their defaults
21  logisticRegr = LogisticRegression(max_iter=100)
22  #logisticRegr = LogisticRegression()
23
24  #do the logistic regression
25  logisticRegr.fit(x_train, y_train)
26  # Use score method to get accuracy of model
27  score = logisticRegr.score(x_test, y_test)
28  print('Percent correct =',score)
```

Line 4 imports the necessary methods. Line 10 applies the default scaling method using `processing.scale` and stores the scaled data in `data_scaled`. Note that you need to use the same scaling on the test data as is done on the training data. In this case, we are scaling the entire set before we split the data so we know that the same scale has been implemented. Line 17 is changed to split the scaled data instead of the original data. Finally, we used trial and error again to see if scaling the data reduced the number of iterations needed to fit the model. Indeed, we could reduce `max_iter` from 2,800 to 100. Thus, scaling greatly reduces the time needed to solve the model. The output follows.

**Output:**
```
Image Data Shape (1797, 64)
Label Data Shape (1797,)
Percent correct = 0.9666666666666667

Process finished with exit code 0
```

The value of `score` calculated in line 27 is achieved by computing predictions for all of the observations in the test set and comparing the predictions to the actual target values that are known. Thus, this model is predicting correctly about 97 % of the time. These images vary in quality and definition. Some are "typed" numbers, while; some are handwritten. To get 97 % correct seems amazing.

Let's generate the confusion matrix as we did before. To do this, we need to include the following lines of code. In the top of the code, we need the following statement:

```
from sklearn import metrics
```

Then, after the model has been fit, we include the following lines.

**Code:**
```
30    predictions = logisticRegr.predict(x_test)
31    print('Number of predictions =',len(predictions))
32    cm = metrics.confusion_matrix(y_test, predictions)
33    print(cm)
```

The program will give us the following confusion matrix.

**Output:**
```
Number of predictions = 450
[[37  0  0  0  0  0  0  0  0  0]
 [ 0 40  0  0  0  0  0  0  2  1]
 [ 0  0 43  1  0  0  0  0  0  0]
 [ 0  0  0 44  0  0  0  0  1  0]
```

```
[ 0    0    0    0  37    0    0    1    0    0]
[ 0    0    0    0    0  46    0    0    0    2]
[ 0    1    0    0    0    0  51    0    0    0]
[ 0    0    0    0    1    0    0  47    0    0]
[ 0    3    1    0    0    0    0    0  44    0]
[ 0    0    0    0    0    0    1    0    0  46]]
```

We can see that the test data included 450 images. From the confusion matrix, we can see that all 37 0's were predicted correctly. Of the forty-three 1's, all but three were predicted correctly. Of those three, two were predicted as 8's and one was predicted as a 9. We can read similar results for all the predictions using the matrix. There were 15 errors among the 450 predictions.

**See Exercises 4–5.**

## 7.3 Neural networks

Our last topic to consider is a very powerful technique for classification known as *neural networks*. Detailed literature on the mathematics underlying neural networks is actually fairly rare. One good resource was found at https://d2l.ai/chapter_preface/index.html, but even that does not fully explain the mathematical process. To be sure, the complexity of neural networks is intense and well beyond our scope here. We will turn our focus to a broader picture of the process and its implementation in Python. While linear and logistic regression have a set of independent (input) variables and a set of dependent (target) variables, neural networks include intermediate variables between the input and target variables. These intermediate variables make up what are called *hidden layers* of the network. When such hidden layers are used, the method becomes a *deep learning* method. A schematic of a small neural network is given in Figure 7.3.

Each circle represents a *node* of the network. The first column of nodes, denoted $I_1$ and $I_2$, comprise the input layer of the network. The middle column, $H_1$, $H_2$, and $H_3$, make up a *hidden layer*, and the last column, $O_1$ and $O_2$, make up the *output layer*. The output layer corresponds to the classes into which we wish to assign the object associated with the inputs. In this network, there are only two classes. In theory we can have any finite number of nodes in any layer, and we could have multiple hidden layers if desired. The arrows (edges) in the network represent the idea that each node of one layer contributes to the value of each node of the next layer. Further, each node would have a weight associated with it to indicate the amount of contribution for the associated node. For example, $w_{11}$ is the weight (coefficient) for $I_1$'s contribution to $H_1$, $w_{12}$ is $I_1$'s contribution to $H_2$, and $w_{ij}$ is the weight assigned to node $i$ when computing node $j$ in the next layer. Finally, each layer after the input layer is assigned a bias variable that will shift the values of the next layer in a way that eases the decision making process later in the

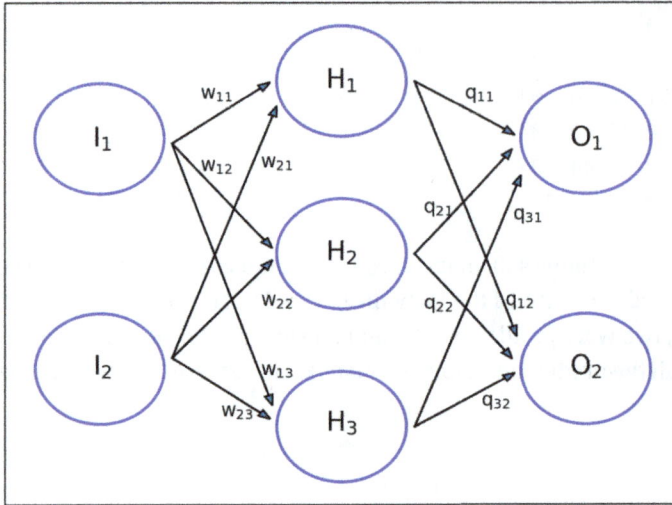

**Figure 7.3:** Representation of a simple neural network.

network. For the network under discussion, the inputs would be processed as follows. We let the lower case letters represent particular values of the associated uppercase node. Thus, $i_1$ is a particular value for $I_1$, $h_2$ a value for $H_2$, and so on. We also assign $b$ as the bias for the hidden layer and $k$ as the bias for the output layer. Then,

$$h_1 = w_{11}i_1 + w_{21}i_2 + b$$
$$h_2 = w_{12}i_1 + w_{22}i_2 + b$$
$$h_3 = w_{13}i_1 + w_{23}i_2 + b.$$

Before using $h_1$, $h_2$, and $h_3$ to compute the output values, an *activation function* is applied. This is similar to what we saw in logistic regression when the logistic function was applied to achieve the final output. Several different activations may be used, but most of them attempt to map the computed values to a number between 0 and 1. A logistic function with 0 as the lower bound and 1 as upper bound is commonly used. The function is called a sigmoid function and is defined as

$$S(x) = \frac{1}{1 + e^{-x}}.$$

Then, once the hidden values have been computed, the activation is applied to each value:

$$a_1 = S(h_1)$$
$$a_2 = S(h_2)$$
$$a_3 = S(h_3).$$

These transformed values are then used to compute the next level:

$$o_1 = q_{11}a_1 + q_{21}a_2 + q_{31}a_3 + k$$
$$o_2 = q_{12}a_1 + q_{22}a_2 + q_{32}a_3 + k.$$

Finally, we apply the activation function again to achieve the final output, $r_1, r_2$.

$$r_1 = S(o_1)$$
$$r_2 = S(o_2).$$

The larger value of $r_1$ and $r_2$ dictates which class is chosen in the decision. The process just described is called *forward propagation*. While it is typical to represent a network as shown in Figure 7.3, this depiction does not explicitly show the activations as data is propagated through the network. A more explicit diagram would be something like that given here.

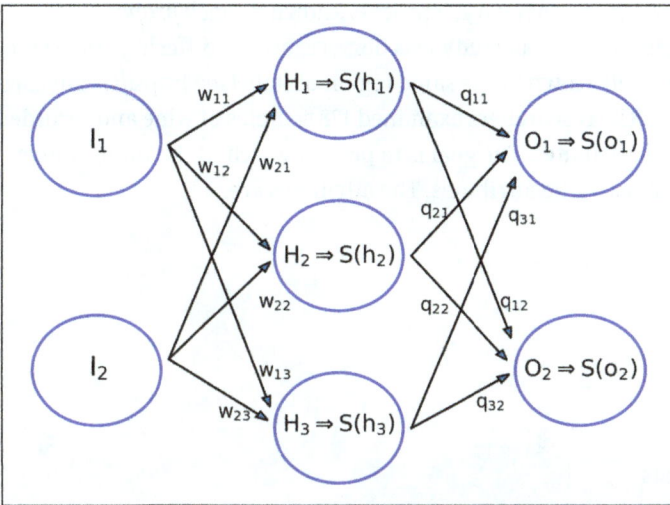

We must have data to fit the network (the more the better), and each data record would include measures of the aspects of interest and the class or category associated with the record. For example, suppose we have data that has characteristics of dogs and cats. Each record of data might include measures like weight, height, color of fur, length of tail, etc. But, each record would also include whether the animal was in the dog class or the cat class. So if $O_1$ represents dog and $O_2$ represents cat, then for a dog we would know that $r_1 = 1$ and $r_2 = 0$. If the animal is a cat, then $r_1 = 0$ and $r_2 = 1$. Now, the values of $r_1$ and $r_2$ computed by the network will likely not be exactly 1 or 0. They will be decimal numbers between 0 and 1. Thus, there is error in the final output. To fit the weights to the input and interior nodes, one must develop a function to indicate the magnitude of the error for each data record. We sum all of the errors to get a final value for the error function. Then, we try to minimize the error function. With so many variables, this minimization is not trivial. Even in the small network depicted, there are 14 variables:

$$w_{11}, \ w_{12}, \ w_{13}, \ w_{21}, \ w_{22}, \ w_{23}, \ b, \ q_{11}, \ q_{12}, \ q_{21}, \ q_{22}, \ q_{31}, \ q_{32}, \ k.$$

This is very complicated and is generally accomplished with a method called *steepest descent* or *gradient descent*. Such methods are beyond the scope of this text, but the idea is this. Initial values for all the variables are assigned (either randomly or by educated estimation). The data is then propagated through the network, and the error is calculated. Using the current values of all the nodes, the weights are adjusted in a way that seeks a lower value for the error. The new weights are applied, and the data is propagated again. The process is repeated until the error function is either unchanging or has reached an acceptable tolerance. The steepest descent algorithm will use the derivatives of the error function to adjust the weights and biases, and these derivatives can be found by working backward through the network, a process known as *back propagation*.

Let's consider an example. A small study was done regarding differing grape cultivars used to produce wine, all grown in the same region of Italy (see https://archive.ics. uci.edu/ml/datasets/wine). The researchers examined 178 samples of wine and recorded measurements for several attributes. The goal is to predict which of the three cultivars was used based on the values of the attributes. The attributes are:
1. Alcohol
2. Malic acid
3. Ash
4. Alcalinity of ash
5. Magnesium
6. Total phenols
7. Flavanoids
8. Nonflavanoid phenols
9. Proanthocyanins
10. Color intensity
11. Hue
12. OD280/OD315 of diluted wines
13. Proline

The file *wine.csv* contains the data for all 178 samples. As the data scientist, we do not have to be experts in wine to construct a useful model. We need to know that there are 13 input variables and 3 classes. Within the data file, the type of cultivar is given as the first field of each record, followed by values for each of the 13 attributes. A sample line (record) is shown here:

2,11.84,2.89,2.23,18,112,1.72,1.32,.43,.95,2.65,.96,2.52,500

Clearly, commas are used to separate the values of the attributes. For this example, we know that this wine was grown with a type 2 cultivar. From a neural network perspective, we need an input layer with 13 nodes and an output layer with 3 nodes. The decision on the number of hidden layers and the number of nodes within each hidden layer is difficult to determine. But once we have the code in Python, we can try different combinations to see if significant gains are made. For our example, we will begin with one hidden layer containing 5 nodes. To construct this network in Python, we begin by importing all of the relevant packages.

**Code:**

```
1   print('importing packages')
2   import numpy as np
3   from sklearn.model_selection import train_test_split
4   import matplotlib.pyplot as plt
5   from sklearn.preprocessing import StandardScaler
6   from sklearn import metrics
7   from sklearn.neural_network import MLPClassifier
8   print('packages imported')
```

Most of these packages have been used previously in the text. Rather than import the entire preprocessing module, we have imported only the standard scaling routines that are included. The code specific to the fitting of the neural network is imported in line 7. Also, print statements are used to inform us as to what the program is doing at each step. These prints act as a kind of status update so that we know the code is indeed proceeding as expected. Once the packages have been imported, the data file is read and the data is stored.

**Code:**

```
9    print('reading data')
10   winedata = np.genfromtxt('wine.csv', delimiter=',')
11   m,n = winedata.shape
12   print('winedata dimensions:',m,n)
13   x = winedata[:,1:]
14   y = winedata[:,0]
```

The data is read by line 10. Lines 11 and 12 determine the size of the matrix that holds the data and displays this information on the screen. Line 13 stores columns 2–14 in a matrix called x, while line 14 stores the first column of the matrix in a variable called y. The x matrix holds the wine attributes, and the y matrix holds the associated classes. As we did in the logistic regression, we should split the data into a training set and a testing set. With such a small data set, we must decide if a split still leaves enough data to effectively build the model, but splitting is, in general, a good practice.

**Code:**

```
15   x_train, x_test, y_train, y_test = \
16       train_test_split(x, y, test_size=0.25,random_state=0)
17
18   print('x_train.shape',x_train.shape)
19   print('y_train.shape',y_train.shape)
20   print('x_test.shape',x_test.shape)
21   print('y_test.shape',y_test.shape)
```

Lines 15 and 16 accomplish the split into a training set and a testing set. The training set contains 75 % of the data while the testing set contains the remaining 25 %. The print statements are used to show the number of records (rows) and attributes (columns) in each set. This enables one to check that the split was achieved as expected. We should also scale the data to reduce the likelihood and magnitude of numerical errors during the fitting of the weights to the data. Recall that we applied scaling in the logistic example in the previous section. In that example, we scaled all of the data before the split. This ensured that all the data received the same transformation. Since the data in our wine example is already split into training and testing sets, we have to go about the scaling in a slightly different way. We must apply the same scaling to the testing data that is applied to the training data. To do this, we "fit" a scaler to the training data and then apply the same scaler to the testing data.

**Code:**

```
22   scaler = StandardScaler()
23   # Fit only to the training data
24   scaler.fit(x_train)
25   # Now apply the transformations to the data:
26   x_train = scaler.transform(x_train)
27   x_test = scaler.transform(x_test)
```

The StandardScaler module is imported in line 5. Line 22 sets up a variable of type StandardScaler. We called the variable scaler, but any valid variable name is acceptable. In line 24, a scaling model is fit to the training data. Then lines 26 and 27 apply the same transformation to both the training and testing data. With the scaling complete, we are prepared to fit the neural network to our data. As with the logistic regression, we use the training data to fit the parameters of the network. The code follows.

**Code:**

```
28   print('fit to neural net')
29   winenetwork = MLPClassifier(hidden_layer_sizes=(10),max_iter=1000)
30   winenetwork.fit(x_train,y_train)
```

A variable called winenetwork of type MLPClassifier is initialized in line 29. For the record, MLP stands for *multi-layer perceptron*. The network is initialized with one hidden layer having ten nodes. Also, the maximum number of iterations for the optimization of the parameters is set to 1,000 (much like we did with the logistic regression). The network is actually constructed in line 30 using the .fit method of the MLPClassifier class. When the entire code is executed, the following output is displayed.

**Output:**

```
importing packages
packages imported
reading data
winedata dimensions: 178 14
x_train.shape (133, 13)
y_train.shape (133,)
x_test.shape (45, 13)
y_test.shape (45,)
fit to neural net

Process finished with exit code 0
```

From the output, we see that the code was executed with no errors (exit code of 0). We can also see that the original data contained 178 records. The split data seems appropriate since the training set has 133 and the testing set has 45 records. These sum to the 178 as expected. The x matrix has 13 columns which correspond with the 13 wine attributes, and the y matrix has only one column that corresponds to the class of cultivar associated with the wine. Hence, it seems that things are working properly, but we have no idea of how well the network will predict classes for the test data. To acquire the prediction associated with a set of input values, we use the .predict method for our winenetwork variable. We can pass the input values for a single observation or for multiple observations. The predictions for all of the testing data are computed in line 31 of the following code.

**Code:**

```
31   predictions = winenetwork.predict(x_test)
32   proport_correct = winenetwork.score(x_test, y_test)
33   print('proportion of correct predictions',proport_correct)
34   # get the confusion matrix
```

```
35    cm = metrics.confusion_matrix(y_test, predictions)
36    print('Confusion Matrix:')
37    print(cm)
```

The .score method compares the predicted values for the inputs with the actual classes associated with the inputs. The method returns the proportion of predictions that are correct. For our wine data, the score is computed in line 32 and displayed in line 33. We should note that, because the weights in the model are initially chosen via a random process, the score could change when the code is executed again (even though the random state for the split was fixed). We could fix the randomness of the weights by assigning a rand_state parameter in line 29, when the MLPClassifier is initialized. Finally, the confusion matrix is also printed in line 35. The output for a particular execution is given next.

**Output:**
```
importing packages
packages imported
reading data
winedata dimensions: 178 14
x_train.shape (133, 13)
y_train.shape (133,)
x_test.shape (45, 13)
y_test.shape (45,)
fit to neural net
proportion of correct predictions 0.9777777777777777
Confusion Matrix:
[[16  0  0]
 [ 1 20  0]
 [ 0  0  8]]


Process finished with exit code 0
```

As before, the confusion matrix gives us more detail on the predictions that were made. For this particular case, there was only one incorrect prediction among the testing set. One observation that was classified as type 1 was predicted as a type 0. All other predictions were correct.

The sklearn package also includes routines to compute particular measures for a given network. One such routine offers a *classification report*. Consider the example confusion matrix from our previous digit recognition model. The matrix and associated classification report are given next.

Confusion Matrix:
```
[[37  0  0  0  0  0  0  0  0  0]
 [ 0 42  0  0  0  0  0  0  1  0]
 [ 0  0 44  0  0  0  0  0  0  0]
 [ 0  0  0 44  0  0  0  0  1  0]
 [ 0  0  0  0 37  0  0  1  0  0]
 [ 0  0  0  0  0 47  0  0  0  1]
 [ 0  1  0  0  0  0 51  0  0  0]
 [ 0  0  0  0  0  0  0 48  0  0]
 [ 0  1  0  1  0  0  0  0 46  0]
 [ 0  0  0  0  0  1  0  1  0 45]]
```

Classification Report:

|  | precision | recall | f1-score | support |
|---|---|---|---|---|
| 0 | 1.00 | 1.00 | 1.00 | 37 |
| 1 | 0.95 | 0.98 | 0.97 | 43 |
| 2 | 1.00 | 1.00 | 1.00 | 44 |
| 3 | 0.98 | 0.98 | 0.98 | 45 |
| 4 | 1.00 | 0.97 | 0.99 | 38 |
| 5 | 0.98 | 0.98 | 0.98 | 48 |
| 6 | 1.00 | 0.98 | 0.99 | 52 |
| 7 | 0.96 | 1.00 | 0.98 | 48 |
| 8 | 0.96 | 0.96 | 0.96 | 48 |
| 9 | 0.98 | 0.96 | 0.97 | 47 |
| | | | | |
| accuracy | | | 0.98 | 450 |
| macro avg | 0.98 | 0.98 | 0.98 | 450 |
| weighted avg | 0.98 | 0.98 | 0.98 | 450 |

The classes for this classifier are given on the left: $0, 1, 2, 3, \ldots, 9$. Some of the measures reported in the classification report deal with the ideas of *false positives* and *false negatives*. A *true positive* classification for $j$ occurs when an object of class $j$ is predicted to be in class $j$. A *false positive* for $j$ occurs when an object of a class other than $j$ is classified as class $j$. The *precision* of a classifier measures the accuracy of positive predictions:

$$\text{precision}(j) = \frac{\text{true positive predictions for } j}{\text{true positive predictions for } j + \text{false positive predictions for } j}$$

In terms of the confusion matrix $CM$,

$$\text{precision}(j) = \frac{CM_{jj}}{\text{Sum of column } j}.$$

Note that, if an object is of class $j$ but is identified as class $i$ ($i \neq j$), then this object is not included in the computation of precision. Such a classification is called a *false negative*. Likewise, a *true negative* occurs if an object not in class $j$ is assigned a class other than $j$.

*Recall* is a measure of how many of the true positives were detected. In this case the false negative classifications are incorporated:

$$recall(j) = \frac{\text{true positives for } j}{\text{true positives for } j + \text{false negatives for } j}$$
$$= \frac{CM_{jj}}{\text{Sum of row } j}.$$

The $f1$ measurement is a weighted average of precision and recall. It is the preferred measure to use when comparing different classifier models:

$$f1 = \frac{2(\text{precision})(\text{recall})}{\text{recall} + \text{precision}}.$$

The *support* for a class is the actual number of occurrences of that class. *Accuracy* is the proportion of predictions that are correct:

$$accuracy = \frac{\text{sum of diagonal of CM}}{\text{sum of all elements in CM}}.$$

The *macro average* is the unweighted average of the measure, and the *weighted average* uses the supports as weights when computing the average of the measure. So the weighted average is given by

$$weighted\ avg = \frac{\sum_{i \text{ in class}} (\text{support}_i)(\text{measure value}_i)}{\sum_{i \text{ in class}} \text{support}_i}.$$

We could write Python functions to compute each of these measures, but sklearn provides a method to generate all of these easily. The report is generated with one line of code.

**Code:**

```
38   print(metrics.classification_report(y_test, predictions))
```

A separate execution of the program including the classification report for our wine network produces the following.

**Output:**
```
importing packages
packages imported
reading data
winedata dimensions: 178 14
x_train.shape (133, 13)
y_train.shape (133,)
x_test.shape (45, 13)
y_test.shape (45,)
fit to neural net
proportion of correct predictions 0.9777777777777777
Confusion Matrix:
[[16  0  0]
 [ 0 20  1]
 [ 0  0  8]]
              precision    recall  f1-score   support

         1.0       1.00      1.00      1.00        16
         2.0       1.00      0.95      0.98        21
         3.0       0.89      1.00      0.94         8

    accuracy                           0.98        45
   macro avg       0.96      0.98      0.97        45
weighted avg       0.98      0.98      0.98        45


Process finished with exit code 0
```

Examination of the confusion matrix shows that $CM(1, 2) = 1$. This reflects a false negative for class 2 and a false positive for class 3. Thus, the precision for class 3 is given by $\frac{8}{0+1+8} \approx 0.89$, while the recall for class 2 is given by $\frac{20}{0+20+1} \approx 0.95$.

**See Exercise 6.**

We close this section with a brief discussion regarding the presentation of the confusion matrix. For the programmer, the appearance of the output of the confusion matrix is not generally of concern. The programmer knows what the matrix represents and how to interpret the entries of the matrix. However, if one is including such information in a presentation of sorts, then a more eloquent display may be more appropriate. There are library routines that create a color-based confusion matrix. One such routine is the .heatmap method found in the *seaborn* package, but we know enough Python now to produce our own meaningful, appealing representation of the confusion matrix. To do this, we can treat the confusion matrix much like we did the digit data from the previous section. Then, we can use the .imshow method to display the matrix. The only code needed is:

```
plt.imshow(cm,cmap ='Blues')
```

Adding this line and rerunning the code produces the following confusion matrix and associated image.

**Output:**
```
Confusion Matrix:
[[16  0  0]
 [ 1 20  0]
 [ 0  0  8]]
```

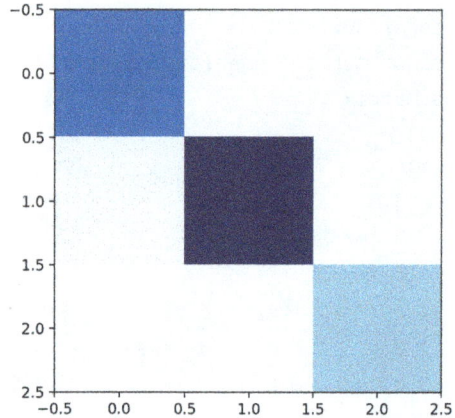

Of course, one may choose any available colormap to be used. The image is nice, but certainly there is some room for improvement. The first thing to notice is that $CM(0,0)$ and $CM(2,2)$ are both the only nonzero entries in their respective rows, but the respective color shading is very different. It would be more appropriate to scale the entries by the row sums (or column sums) so that shadings are more representative of the accuracies in the network predictions. Also, the axis scalings are not meaningful. We should include only the class labels. We can accomplish both of these goals with the following code.

**Code:**
```
39    rowsums = np.sum(cm,0)
40    scaledcm = cm/rowsums
41    plt.imshow(scaledcm,cmap ='Blues',alpha=0.75)
42    plt.xticks(np.arange(0,3,1),['1','2','3'])
43    plt.yticks(np.arange(0,3,1),['1','2','3'])
44    plt.show()
```

Line 39 computes the row sums for all the rows in cm. If we were to replace the 0 with a 1, column sums would be computed instead of row sums. Line 40 divides all the entries in cm by the respective row sum. The image is produced in line 41 (the alpha parameter allows us to set the transparency of the color). Finally, lines 42–43 define the axis tick-mark locations and the labels that are to be used. When executed, the following confusion matrix and image are created.

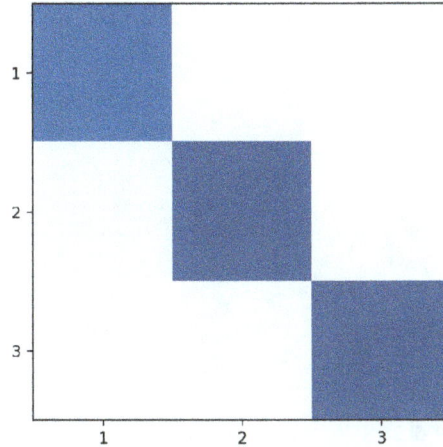

**Output:**
```
Confusion Matrix:
[[16  0  0]
 [ 1 20  0]
 [ 0  0  8]]
```

Finally, it would be helpful if the actual entries of the matrix were shown. We can write text on a graph in matplotlib by using the .text method. The method requires the *x* and *y* coordinates on which text is to be displayed as well as the actual text to be displayed. Since we wish to display the contents of cm, we will use a nested loop system and a little trial and error to determine the locations for the text. The looping structure is added as follows.

**Code:**
```
39  rowsums = np.sum(cm,0)
40  scaledcm = cm/rowsums
41  plt.imshow(scaledcm,cmap ='Blues',alpha=0.75)
42  plt.xticks(np.arange(0,3,1),['1','2','3'])
43  plt.yticks(np.arange(0,3,1),['1','2','3'])
44  for i in range(3):
45      for j in range(3):
46          plt.text(i-.1,j+.05,str(cm[i,j]))
47  plt.show()
```

The end result is illustrated in the next graph.

**Output:**
```
Confusion Matrix:
[[16  0  0]
 [ 1 20  0]
 [ 0  0  8]]
```

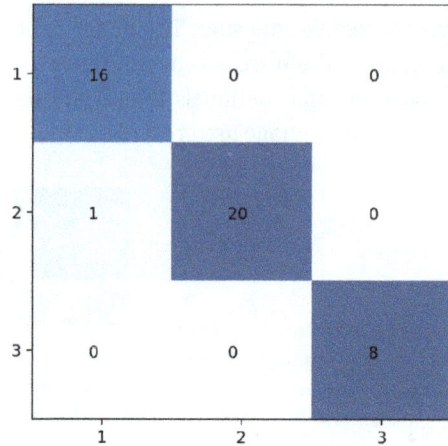

See Exercise 7.

## 7.4 Exercises

1. Use the data from the teen birth example find the linear regression line that predicts births to 18–19-year-old women (instead of 15–17-year-olds) based on poverty rate.
   (a) Create a scatter plot using poverty as the explanatory variable and births as the response variable.
   (b) Plot the regression line on the same plot as the scatter plot.
   (c) Find the coefficient of determination and interpret its value.

2. Efron et al. (2004, in Annals of Statistics) collected data on 442 diabetic patients. The data is contained in *diabetes.txt*. The first four fields of the data are: age, sex (1 = male, 2 = female), body mass index (BMI), and blood pressure (BP). The right-hand column is an indication of disease progression.
   (a) Use this data to find the linear regression model to predict disease progression using age, sex, BMI, and BP as the explanatory variables.
   (b) Find the coefficient of determination and the correlation coefficient.
   (c) Based on the values of the $R^2$ and $r$, comment on the strength of the linear model.
   (d) For a woman who is 45-years-old with BMI of 30 and BP of 112, what is the predicted disease progression?

3. Use the diabetes data from the previous problem.
   (a) Find separate regression models for males and females, using only age, BMI, and blood pressure as explanatory variables.
   (b) Find the $R^2$ values associated with males and females. Comment on the results.

4. Revise the logistic regression for the digits example as follows:

(a) Instead of using all 64 entries in the image matrix, only use the odd numbered rows of the image matrix. Thus, there would be a total of 32 explanatory variables instead of 64.

(b) Build the logistic-regression model for the revised data and compare the prediction accuracy to the example model.

5. Using the diabetes data from Exercise 2, assign classes to the disease progression field (Y) according to the following table.

| Disease Progression | Class |
|---|---|
| 0–50 | 0 |
| 50–100 | 1 |
| 100–150 | 2 |
| 150–200 | 3 |
| above 200 | 4 |

(a) Construct a logistic regression model using age, sex, BMI, and BP as the explanatory variables and the classes described as the target values. Use the entire data set as training data (and the testing data). To avoid warning messages, you may have to increase the `max_iter` parameter in the regression variable.

(b) How accurately are the classes predicted for the given data?

(c) Compare the results from the logistic regression with those from the previous linear regression.

6. Recall the logistic regression model to recognize digits in Section 7.2.

(a) Modify the logistic model to create a neural network model with two hidden layers. The first hidden layer should have 60 nodes and the second hidden layer should have 30 nodes.

(b) Produce a classification report for the resulting network using the test data.

(c) Compare the results of the logistic model and neural network mode.

(d) Add code to produce an appealing display of the confusion matrix for the digit-recognition model. You may modify the code developed in the text or use the `.heatmap` method in the *seaborn* package.

7. Moody et al. collected data regarding heart conditions. The data includes a large number of ECG observations included in the files, *mitbih_test.csv* and *mitbih_train.csv*. (Moody GB, Mark RG). The impact of the MIT-BIH Arrhythmia Database. IEEE Eng in Med and Biol 20(3):45–50 (May–June 2001). (PMID: 11446209). The data has already been split (as the file names indicate) into a training set and a testing set.

(a) Each ECG contains 187 amplitude values. The 188th column is the class for the ECG. Use the training set to find and plot one ECG associated with each class. Include a legend on the plot. The y values are the amplitudes for the ECG. You can let the x values be $[0, 1, 2, \ldots 186]$.

(b) Construct a neural network model to predict the classification of the heart condition indicated by an ECG.

(c) Generate predictions for the testing data set.

(d) Display the classification report.

(e) Display an appealing confusion matrix.

# A Python code

## A.1 Chapter 2 code

```
# Chapter 2:
# 2.1,Arithmetic

# Addition, Subtraction, Multiplication, Division

3+5
3-5
3*5
5/3

# powers
print(5**3)

# printing
print('3+5 =',3+5)
print('The value of 2/3 to four decimal places is {:.4f}. '.format(2/3))
print('The radius is {}, and the area is {:.3f}.'.format(4.0, 3.14*4**2))
print('The radius is {:.2f}, and the area is {:.5f}.'.format(4.0, 3.14*4**2))
```

## A.2 Chapter 3 code

```
# exponentials
# importing packages, math package
import math
print(math.e)

# two representations of the same exponential value
print(math.e**3)
print(math.exp(3))

# can also be done in numpy
import numpy
print(numpy.exp(3))

# trig functions
#print a table of trig values
import numpy
print("angle |{:>5} |{:>5} |{:>5} |{:>5} |{:>5} |".format(\
```

https://doi.org/10.1515/9783110776645-008

```python
        '0','\u03C0/6','\u03C0/4','\u03C0/3','\u03C0/2'))
print('---------------------------------------')
print("cos(x)|{:.4f}|{:.4f}|{:.4f}|{:.4f}|{:.4f}|".format(\
    numpy.cos(0),numpy.cos(numpy.pi/6),numpy.cos(numpy.pi/4),\
    numpy.cos(numpy.pi/3),numpy.cos(numpy.pi/2)))
print("sin(x)|{:.4f}|{:.4f}|{:.4f}|{:.4f}|{:.4f}|".format(\
    numpy.sin(0),numpy.sin(numpy.pi/6),numpy.sin(numpy.pi/4),\
    numpy.sin(numpy.pi/3),numpy.sin(numpy.pi/2)))
# =========================================================

# variable types
r = 4
print(type(r))
q = 3
print(type(q))
a = r/q
print(a)
print(type(a))

# strings
x = 'Will Miles'
print(x)

fname = 'Will'
lname = 'Miles'
name = fname+lname
print(name)

name = fname + ' ' + name
print(name)

fname = 'Will'
name = 4*fname
print(name)

# accessing and splicing strings
coursename = 'Scientific Computing'
print(coursename[3])
print(coursename[0:10])
print(coursename[11:])

# find in string
```

```python
coursename = 'Scientific Computing'
a = coursename.find('Comp')
print(a)
b = coursename.find('not')
print(b)

#use upper case to ignore case sensitivity
coursename = 'Scientific Computing'
# convert string to upper case
Ucourse = coursename.upper()
print(Ucourse, coursename)
#search for the uppercase of 'comp'
a = Ucourse.find('comp'.upper())
print(a)
# ===========================================================

# defining functions
def f(x):
    y = 3.0*x**2-2.0*x+1.0
    return y

y = f(3)
print('f(3)=',y)
# ===========================================================

# input from the keyboard
import numpy as np
radius_str = input('Enter the radius: ')
#convert the radius to a floating point value
radius = float(radius_str)
#compute the area, A = pi*r^2
area = np.pi*radius**2
print('The area of a circle with radius {:.3f} cm is {:.3f}\
  square cm'.format(radius, area))
# ===========================================================

#Python lists
x = [1,2,3,4]
print('x=',x)
y = [1,'a',3.14,'c','will']
print('y=',y)
# ===========================================================
```

```
# graphing
import matplotlib.pyplot as plt
x = [0,1,2,3,4,5]
y = [0,1,4,9,16,25]
plt.plot(x,y,'b*-',label='f(x)=x^2')
plt.xlabel('x-axis')
plt.ylabel('y-axis')
plt.title('Big Title')
plt.grid()
plt.show()
# ----------------------------------------------------------

import matplotlib.pyplot as plt
x = [0,1,2,3,4,5]
y = [0,1,4,9,16,25]
z = [0,2,4,6,8,10]
plt.plot(x,y,'b*-',label='f(x)=x^2')
plt.xlabel('x-axis')
plt.ylabel('y-axis')
plt.title('Big Title')
plt.grid()
plt.plot(x,z,'b--',label="f'(x)=2x",c='0.45')
plt.legend()
plt.show()
# ----------------------------------------------------------

import numpy as np
def f(x):
    y = x**2
    return y

x = np.arange(0,5.1,0.1)
print('x=',x)
y = f(x)
print('y=',y)
# ----------------------------------------------------------

import numpy as np
import matplotlib.pyplot as plt
np.set_printoptions(precision=3,suppress=1,floatmode='fixed')
def f(x):
```

```
    y = x**2
    return y

x = np.arange(0,5.1,0.1)
y = f(x)
plt.plot(x,y)
plt.xlabel('x-axis')
plt.ylabel('y-axis')
plt.title('f(x) = x^2')
plt.grid()
plt.show()
# ----------------------------------------------------------

import numpy as np
import matplotlib.pyplot as plt
np.set_printoptions(precision=3,suppress=1,floatmode='fixed')
def f(x):
    y = x**2
    return y

x = np.arange(0,5.1,0.1)
y = f(x)
z = np.sin(x)
plt.plot(x,y)
plt.grid()
plt.figure()
plt.plot(x,z)
plt.grid()
plt.show()
# ==========================================================

#our first if statement
#first get an x value
x = float(input('Enter an x value: '))
#logic of the piecewise function
if x<=0:             #the condition is x<=0
    y = x**2         #condition is true
else:            #otherwise
    y = x+1      #condition is false
print('f({}) = {}'.format(x,y))
# ----------------------------------------------------------
```

```python
# vectorizing functions
import numpy as np
import matplotlib.pyplot as plt
np.set_printoptions(precision=3,suppress=1,floatmode='fixed')
# define a piecewise function using if statements
# in this example, we have named the function pw
def pw(x):
    #logic of the peicewise function
    if x<=0:    #the condition is x<=0
        y = x**2    #condition is true
    else:       #otherwise
        y = x+1 #condition is false
    return y
vpw = np.vectorize(pw)
a = -2
b = 2
n = 100
dx = (b-a)/n
x = np.arange(a,b+dx,dx)
y = vpw(x)
plt.plot(x,y)
plt.show()
# ----------------------------------------------------------

import numpy as np
import matplotlib.pyplot as plt
np.set_printoptions(precision=3,suppress=1,floatmode='fixed')
# define a piecewise function using if statements
# in this example, we have named the function pw
def pw(x):
    #logic of the peicewise function
    if x<=0:    #the condition is x<=0
        y = x**2    #condition is true
    else:       #otherwise
        y = x+1 #condition is false
    return y
vpw = np.vectorize(pw)
#set up the list for -2<x<=0
a = -2
b = 0
n = 50
dx = (b-a)/n
```

```
x = np.arange(a,b+dx,dx)
y = vpw(x)
plt.plot(x,y,'b')
#now do the second section of the function
a = 0
b = 2
n = 50
dx = (b-a)/n
# in this list we want to exclude the left endpoint at 0
# so we will use a start value that is slightly larger than 0
x = np.arange(a+dx,b+dx,dx)    #note, this includes 2
y = vpw(x)
plt.plot(x,y,'b')
plt.grid()
plt.show()
# ----------------------------------------------------------

import numpy as np
import matplotlib.pyplot as plt
np.set_printoptions(precision=3,suppress=1,floatmode='fixed')
# define a piecewise function using if statements
# in this example, we have named the function pw
def pw(x):
    #logic of the peicewise function
    if x<=0:     #the condition is x<=0
        y = x**2     #condition is true
    else:        #otherwise
        y = x+1 #condition is false
    return y
vpw = np.vectorize(pw)
#set up the list for -2<x<=0
a = -2
b = 0
n = 50
dx = (b-a)/n
x = np.arange(a,b+dx,dx)     #last element may be very slightly above zero
x[n] = b                     #assign the last element to be zero
y = vpw(x)
plt.plot(x,y,'b')
#now do the second section of the function
a = 0
b = 2
```

```
n = 50
dx = (b-a)/n
# in this list we want to exclude the left endpoint at 0
# so we will use a start value that is slightly larger than 0
x = np.arange(a+dx,b+dx,dx)     #note, this includes 2
y = vpw(x)
plt.plot(x,y,'b')
plt.grid()
#plt.plot(0,0,'b.',markersize=11)
#plt.plot(0,1,'b.',fillstyle='none',markersize=11)
plt.show()
# ----------------------------------------------------------
```

## A.3 Chapter 4 code

```
# matrices
import numpy as np
#create the matrix
A = np.array([[1,2,3,4],[-2,4,-3,5],[-1,3,-3,4]])
print('A = ')
print(A)
#access the second row, third column
a23 = A[1,2]
print('The value in the second row, third column is ',a23)
#find the size of the matrix
m,n = np.shape(A)
print('rows = {}. cols = {}'.format(m,n))
#get the third row of the matrix
Arow3 = A[2,:]
print('The third row is ',Arow3)
#get the second column of the matrix
Acol2 = A[:,1]
print('The second column is ',Acol2)
# ----------------------------------------------------------

import numpy as np
#create the matrix
A = np.array([[1,2,3,4],[-2,4,-3,5],[-1,3,-3,4]])
print('A = ')
print(A)
#show that matrices are immutable
```

```
B = A
#change B
B[0,0] = 2
#show that A was also changed.
print('A = ')
print(A)
# -----------------------------------------------------------

import numpy as np
#create the matrix
A = np.array([[1,2,3,4],[-2,4,-3,5],[-1,3,-3,4]])
print('Original A = ')
print(A)
#show that matrices are immutable
B = A.copy()
#change B
B[0,0] = 2
#show that A was also changed.
print('A after B has been changed =')
print(A)
print('B after the change =')
print(B)
# -----------------------------------------------------------

# matrix arithmetic
import numpy as np
np.set_printoptions(precision=3,suppress=1,floatmode='fixed')
#create the matrix
A = np.array([[1,2],[3,4]])
B = np.array([[-1,3],[2,-5]])
print('A =')
print(A)
print('B =')
print(B)
#add two matrices
print('A+B = ')
print(A+B)
#subtract B from A
print('A-B = ')
print(A-B)
#component-wise multiplication
print('A*B = ')
```

```
print(A*B)
#component-wise division
print('A/B = ')
print(A*B)
#scalar multiplication
#multiply A by 3
print('3A = ')
print(3*A)
# ---------------------------------------------------------

import numpy as np
np.set_printoptions(precision=3,suppress=1,floatmode='fixed')
#create the matrix
A = np.array([[1,2,3],[-1,2,-3]])
B = np.array([[1,2],[0,4],[-3,2]])
print('A =')
print(A)
print('B =')
print(B)
#standard matrix multiplication
C = np.dot(A,B)
print('AB =')
print(C)
# =========================================================

# identity matrix
import numpy as np
np.set_printoptions(precision=3,suppress=1,floatmode='fixed')
#create the matrix
A = np.array([[1,2,3],[-1,2,-3]])
I3 = np.array([[1,0,0],[0,1,0],[0,0,1]])
I2 = np.array([[1,0],[0,1]])
print('A =')
print(A)
print('I3 =')
print(I3)
print('I2 =')
print(I2)
#standard matrix multiplication
C = np.dot(A,I3)
print('(A)(I3) =')
print(C)
```

```
D = np.dot(I2,A)
print('(I3)(A) =')
print(D)
#print('(A)(I2) =')
#print(np.dot(A,I2))
# ----------------------------------------------------------

import numpy as np
#create the matrix
A = np.array([[1,2,3],[-1,2,-3],[0,2,5]])
print('A =')
print(A)
#find the inverse of A
A_inv = np.linalg.inv(A)
print('A_Inverse =')
print(A_inv)
#confirm the inverse
print('AA_inv = ')
print(np.dot(A,A_inv))
print('A_invA = ')
print(np.dot(A_inv,A))
# ----------------------------------------------------------

# round numbers in a matrix
#confirm the inverse
print('AA_inv = ')
AA_inv = np.dot(A,A_inv)
#round the entries to 3 decimal places when printing
print(AA_inv.round(3))
print('A_invA = ')
A_invA=np.dot(A_inv,A)
print(A_invA.round(3))
# ==========================================================

# linear systems
# row operations
import numpy as np
#np.set_printoptions(precision=3,suppress=1,floatmode='fixed')
#create the matrix
A = np.array([[1,1,1,6],[2,-3,1,-1],[1,2,-3,-4]])
print('A =')
print(A)
```

```
#perform row operations to achieve the goal matrix
#-2R1+R2-->R2
print('-2R1+R2-->R2')
A[1,:] = -2*A[0,:]+A[1,:]
print('A =')
print(A)
#-R1+R3-->R3
print('-R1+R3-->R3')
A[2,:] = -1*A[0,:]+A[2,:]
print('A =')
print(A)
print('(-1/5)R2-->R2')
A[1,:] = (-1/5.0)*A[1,:]
print('A =')
print(A)
print('-1R2+R1-->R1')
print('-1R2+R3-->R3')
A[0,:] = -1*A[1,:]+A[0,:]
A[2,:] = -1*A[1,:]+A[2,:]
print('A =')
print(A)
print('(-1/4)R3-->R3')
A[2,:] = (-1/4.0)*A[2,:]
print('A =')
print(A)
print('-1R3+R1-->R1')
A[0,:] = -1*A[2,:]+A[0,:]
print('A =')
print(A)
# --------------------------------------------------------

# use inverse to solve system
import numpy as np
#create the coefficient matrix
A = np.array([[1,1,1],[2,-3,1],[1,2,-3]])
#create the right hand side column vector
B = np.array([[6],[-1],[-4]])
print('A =')
print(A)
print('B =')
print(B)
#get the inverse of A
```

```python
AInv = np.linalg.inv(A)
#multiply the inverse of A by B
X = np.dot(AInv,B)
print('X =')
print(X)
# -----------------------------------------------------------

# solve system using .solve
import numpy as np
#np.set_printoptions(precision=3,suppress=1,floatmode='fixed')
#create the matrix
A = np.array([[1,1,1],[2,-3,1],[1,2,-3]])
B = np.array([[6],[-1],[-4]])
print('A =')
print(A)
print('B =')
print(B)
X = np.linalg.solve(A,B)
print('X =')
print(X)
```

## A.4 Chapter 5 code

```python
# bisection background
import numpy as np

def f(x):
    y = x**4/10 -2*x**2 + -x-3*np.sin(x) + 5
    return(y)

a = 0
b = 2
print('f({:.4f}) = {:.4f}'.format(a,f(a)))
print('f({:.4f}) = {:.4f}'.format(b,f(b)))
x = (a+b)/2.0
print('f({:.4f}) = {:.4f}'.format(x,f(x)))
# -----------------------------------------------------------

# using while statement in bisection algorithm
import numpy as np
```

```python
def f(x):
    y = x**4/10 -2*x**2 + -x-3*np.sin(x) + 5
    return(y)


tol = 0.0001
a = 0
b = 2
#we will do the first iterate before our while loop starts so that we
#have a value to test against the tolerance
x = (a+b)/2
while np.abs(f(x))>tol:
    print('a={:.5f}  f(a)={:.5f},    b={:.5f}  f(b)={:.5f},   \
        x={:.5f}  f(x)={:.5f}'.format(a,f(a),b,f(b),x,f(x)))
    #now decide whether we replace a or b with x
    if f(a)*f(x) < 0:
        #root is between a and x so replace b
        b = x
    elif f(b)*f(x)<0:
        #root is between b and x so replace a
        a = x
    else:
        # in this case, f(x) must be 0 and we have found the root
        # so we will know the root value is x and we can end the loop
        break
    #recompute the approximation
    x = (a+b)/2
print('final x =',x)
print('final f(x) =',f(x))
# ---------------------------------------------------------


# bisection as a function
def bisect(f,a,b,tol):
    #we will do the first iterate before our while loop starts so that we
    #have a value to test against the tolerance
    x = (a+b)/2
    while np.abs(f(x))>tol:
        print('a={:.5f}  f(a)={:.5f},    b={:.5f}  f(b)={:.5f}, \
            x={:.5f}  f(x)={:.5f}'.format(a,f(a),b,f(b),x,f(x)))
        #now decide whether we replace a or b with x
        if f(a)*f(x) < 0:
            #root is between a and x so replace b
            b = x
```

```python
        elif f(b)*f(x)<0:
            #root is between b and x so replace a
            a = x
        else:
            # in this case, f(x) must be 0 and we have found the root
            # so we will know the root value is x and we can end the loop
            break
        #recompute the approximation
        x = (a+b)/2
    return x
# ------------------------------------------------------------

# using the bisection function
def shifted_exp(x):
    y = np.exp(x) - 3
    return y

tol = 0.0001
a = 1
b = 2
x = bisect(shifted_exp,a,b,tol)
print('final x =',x)
print('final f(x) =',shifted_exp(x))
# ============================================================

# for loop to do sum
import numpy as np
n = 5
t = np.arange(1,n+1,1)
print(t)
S = 0
for k in t:
    S = S+1.0/k
print('S =',S)
# ------------------------------------------------------------


# Euler's method
import numpy as np
import matplotlib.pyplot as plt

#solve dy/dt = (y-t)/2
```

```python
#where when t=0, y=1

def rhs(t,y):
    m = (y-t)/2
    return m

#initial t value
a = 0
#final t value
b = 5
#number of intervals
n = 10
#delta t
dt = (b-a)/n

#create a vector of t-values
t = np.arange(a,b+dt,dt)
#create space for the y-values
y = np.zeros(n+1)
#create a list of indices
i = np.arange(1,n+1,1)
#we know the inital value of y to be 1
y[0] = 1
for k in i:
    #compute the Euler approximation
    #use the right hand side function to get the slope of the tangent line
    m = rhs(t[k-1],y[k-1])
    #get the next approximation
    y[k] = m*dt+y[k-1]
#plot the solution
plt.plot(t,y)
plt.autoscale(enable=True, axis='x', tight=True)
plt.xlabel('t')
plt.ylabel('y')
plt.grid()
plt.show()
# -----------------------------------------------------------

# Euler's method for systems
import numpy as np
import matplotlib.pyplot as plt
```

```python
def rhs(t,yvec):
    dy = np.zeros(2)
    dy[0] = yvec[1]**2-yvec[0]
    dy[1] = (yvec[1]-yvec[0])/2
    return dy

#initial t value
a = 0
#final t value
b = 20
#number of intervals
n = 500
#delta t
dt = (b-a)/n

#create a vector of t-values
t = np.arange(a,b+dt,dt)
#create space for the y-values
y = np.zeros((n+1,2))
#create a list of indices
i = np.arange(1,n+1,1)
y[0,0] = 1.5
y[0,1] = 1
for k in i:
    #compute the Euler approximation
    #use the right hand side function to get the slope of the tangent line
    dy = rhs(t[k-1],y[k-1,:])
    #get the next approximation
    y[k,:] = dy*dt+y[k-1,:]
#plot the approximations
plt.plot(t,y[:,0])
plt.plot(t,y[:,1])
#plot true solution
plt.autoscale(enable=True, axis='x', tight=True)
plt.xlabel('t')
plt.grid()
plt.legend(['x(t)','y(t)'])
plt.show()
# ----------------------------------------------------------

# phase portrait
plt.figure()
```

```python
plt.plot(y[:,0],y[:,1])
head = 1
tail = 0
w = 55
dx = y[head,0]-y[tail,0]
dy = y[head,1]-y[tail,1]
plt.arrow(y[head,0],y[head,1],dx,dy,width=.004)
numarrows = int((n-head)/w)
for i in range(4):
    head = head + w
    tail = tail + w
    dx = y[head,0]-y[tail,0]
    dy = y[head,1]-y[tail,1]
    plt.arrow(y[head,0],y[head,1],dx,dy,width=.004)
plt.xlabel('x(t)')
plt.ylabel('y(t)')
plt.title('Phase Portrait: IC = (1.5,1)')
plt.grid()
plt.show()
# ----------------------------------------------------------

# Euler method for second order equation
import numpy as np
import matplotlib.pyplot as plt

def rhs(t,yvec):
    mu = 1
    dy = np.zeros(2)
    dy[0] = mu*(1-yvec[1]**2)*yvec[0]-yvec[1]
    dy[1] = yvec[0]
    return dy

#initial t value
a = 0
#final t value
b = 20
#number of intervals
n = 500
#delta t
dt = (b-a)/n

#create a vector of t-values
```

```
t = np.arange(a,b+dt,dt)
#create space for the y-values
y = np.zeros((n+1,2))
#create a list of indices
i = np.arange(1,n+1,1)
y[0,0] = 1
y[0,1] = 1
for k in i:
    #compute the Euler approximation
    #use the right hand side function to get the slope of the tangent line
    dy = rhs(t[k-1],y[k-1,:])
    #get the next approximation
    y[k,:] = dy*dt+y[k-1,:]
#plot the approximations
plt.plot(t,y[:,0])
plt.plot(t,y[:,1])
#plot true solution
plt.autoscale(enable=True, axis='x', tight=True)
plt.xlabel('t')
plt.grid()
plt.legend(['x(t)','y(t)'])
plt.figure()
plt.plot(y[:,0],y[:,1])
head = 1
tail = 0
w = int(n/12)
dx = y[head,0]-y[tail,0]
dy = y[head,1]-y[tail,1]
plt.arrow(y[head,0],y[head,1],dx,dy,width=.01)
numarrows = int((n-head)/w)
for i in range(4):
    head = head + w
    tail = tail + w
    dx = y[head,0]-y[tail,0]
    dy = y[head,1]-y[tail,1]
    plt.arrow(y[head,0],y[head,1],dx,dy,width=.025)
plt.xlabel('x(t)')
plt.ylabel('y(t)')
plt.title('Phase Portrait: IC = (1,1)')
plt.grid()
plt.show()
# ===========================================================
```

```python
# interpolation
import numpy as np

# x is the new input value, t is the vector of x-values
# y is the vector of y-values
def interp(x,t,y):
    n = len(t)
    startindex = 0
    # find the indices between which the new x value lies
    while t[startindex]<=x:
        startindex = startindex + 1
    startindex = startindex - 1
    endindex = startindex +1
    # slope for interpolation
    m = (y[endindex]-y[startindex])/(t[endindex]-t[startindex])
    # compute approximation using point slope form
    y_of_x = m*(x-t[startindex])+y[startindex]
    return y_of_x


t = np.array([0.000, 0.040, 0.080, 0.120, 0.160,\
            0.200, 0.240, 0.280, 0.320, 0.360])
y = np.array([1.000, 1.040, 1.078, 1.115, 1.150,\
            1.182, 1.212, 1.240, 1.266, 1.289])

#approximate y(0.0732)
x = 0.0732
yinterp = interp(x,t,y)
print('y({}) = {:.5f}'.format(x,yinterp))
```

## A.5  Chapter 6 code

```python
# file handling
climate_file = open('brazilclimate.csv','r')
count = len(climate_file.readlines())
print('Number of lines in the file is',count)
climate_file.close()
# -----------------------------------------------------------

# splitting strings and cleaning data
```

```
climate_file = open('smallclimate.csv','r')
#we know the first line of this file contains headers of the columns
record = climate_file.readline()
#set up a counter to know what line we are on
count = 1
#set up a counter to count number of nonzero values
num_non0 = 0
# read the second line (which is the first line containing actual data)
record = climate_file.readline()
#read lines until you reach a blank line, then assume you are done
while record != "":
    record = climate_file.readline() #move to bottom of the loop
    #split the record into its separate columns
    record_vector = record.split(',')
    #prcp is the 15th column.  in Python, that is index 14
    #convert from a string to value
    if record_vector[14] == '':
        prcp = 0
    else:
        prcp = float(record_vector[14])
        if prcp != 0.0:
            num_non0 = num_non0 + 1
            print('prcp = :.4f in record #.'.format(prcp,count))
    count = count + 1
    record = climate_file.readline()
print('The number of nonzero values is .'.format(num_non0))
print('There were  records processed.'.format(count))
climate_file.close()



# -----------------------------------------------------------
# data cleaning
# open the climate file for reading
climate_file = open('brazilclimate.csv','r')
# open the new (output) file for writing. If the file does not exist,
# it is created.
temperature_file = open('tempstudy.csv','w')
# write the headers to the output file.  the \n is a next line indicator
headerline = 'ID,lat,long,year,month,day,hour,precip,temp\n'
temperature_file.write(headerline)
# we know the first line of the input file contains headers of the columns
record = climate_file.readline()
```

```
# set up a counter to know what line of the input file we are on
count = 1
# set up counters to count the number of records that are corrected and
# the number of records that are discarded
corrected_recs = 0
discarded_recs = 0
# get the first non-header line (this is the first line with actual data)
record = climate_file.readline()
# read lines until you reach a blank line, then assume you are done
while record != "":
    # initialize a variable (discard flag) to indicate whether
    # the record should be discarded
    # 0 = keep the record (write to the new file),
    # -1 = discard the record (do not write to the new file)
    # discarded records are not deleted from original file
    # initialize the flag to "keep"
    discardflag = 0
    # this just lets me know that the program is progressing through the file
    # by printing to the screen every millionth record
    if count%1000000 == 0:
        print(count)
    # split the input record into its separate columns
    record_vector = record.split(',')
    # check station ID, year, month, temperature for blanks the \ allows us to
    # continue the code line to the next line for readability
    if record_vector[0] == '' or record_vector[10] == '' \
            or record_vector[11] == '' or record_vector[21] == '':
        # if any of them is blank, set the discard flag and increment the
        # the discarded records counter
        discardflag = -1
        discarded_recs = discarded_recs + 1
    # see if there are any blank lat/long values that have a station id
    if record_vector[3] == '' or record_vector[3]=='':
        if record_vector[0] != '':
            print('need to look up station id')
        else:
            # if the lat and long and id are all blank, discard the record
            discardflag = -1
    # now replace blank precip values with 0.  only need to do this in the new
    # file
    if record_vector[14] == '':
        record_vector[14] = '0'
```

```python
        # increment corrected records counter
        corrected_recs = corrected_recs + 1
    # build and write the record to the output file
    if discardflag == 0:
        # temp_record holds the record to be written
        # the + will just concatenate the strings.  the \ is a line
        # continuation character
        # we are building a string with all the desired
        # fields separated by commas
        temp_record = record_vector[0]+','+record_vector[3]+','\
        +record_vector[4]+','+record_vector[10]+','+record_vector[11]+','\
        +record_vector[12]+','+record_vector[13]+','+record_vector[14]+','\
        +record_vector[21]+'\n'
        # write to the output file
        temperature_file.write(temp_record)
    # read the next line of input
    record = climate_file.readline()
    count = count+1
    # go back to the top of the loop
# the loop is complete
# close the files
climate_file.close()
temperature_file.close()
# print the counts
print('Number of discarded records:',discarded_recs)
print('Number of corrected records:',corrected_recs)
# ---------------------------------------------------------

# computing averages of categories
import numpy as np
import matplotlib.pyplot as plt
np.set_printoptions(precision=3,suppress=1,floatmode='fixed')

# open the temp study file for reading
temperature_file = open('tempstudy.csv','r')
# we know the first line of this file contains headers of the columns
record = temperature_file.readline()
# set up a matrix to hold what we need
# we need a row for each month and columns for
# total precip, total temperature, and number of observations
tempsummary = np.zeros((12,3))
# read lines until you reach a blank line, then assume you are done
```

```python
temprec = temperature_file.readline()
count = 0
while temprec != '':
    if count%1000000 == 0:
        print(count)
    # split the record into fields
    tempvec = temprec.split(',')
    # get the month, precipitation, and temperature for this observation
    mth = int(tempvec[4])
    precip = float(tempvec[7])
    temp = float(tempvec[8])
    tempsummary[mth-1,0] = tempsummary[mth-1,0] + precip
    tempsummary[mth-1,1] = tempsummary[mth-1,1] + temp
    tempsummary[mth-1,2] = tempsummary[mth-1,2] + 1
    temprec = temperature_file.readline()
    count = count + 1
temperature_file.close()
tempsummary[:,0] = tempsummary[:,0]/tempsummary[:,2]
tempsummary[:,1] = tempsummary[:,1]/tempsummary[:,2]
print(tempsummary)
# graph results
m = np.arange(1,13,1)
plt.plot(m,tempsummary[:,0])
plt.xlabel('month')
plt.ylabel('precip')
plt.grid()
plt.show()
# write info to file
tempsum_file = open('tempsumm.csv','w')
for i in range(12):
    summrec = str(tempsummary[i,0])+','+str(tempsummary[i,1])+',\
                '+str(tempsummary[i,2])+'\n'
    tempsum_file.write(summrec)
tempsum_file.close()
# ---------------------------------------------------------

# use summary file to produce graphs
import numpy as np
import matplotlib.pyplot as plt
np.set_printoptions(precision=3,suppress=1,floatmode='fixed')
tempsumm_file = open('tempsumm.csv','r')
# set up a matrix to hold what we need
```

```
# we need a row for each month and columns for
# total preip, total temperature, and number of observations
tempsummary = np.zeros((12,3))
# read each line and fill the matrix
for i in range(12):
    summrec = tempsumm_file.readline()
    summvec = summrec.split(',')
    for j in range(3):
        tempsummary[i,j] = summvec[j]
m = np.arange(1,13,1) # a list of month values
plt.plot(m,tempsummary[:,0])
plt.xlabel('month')
plt.ylabel('precip')
plt.grid()
plt.show()
# ---------------------------------------------------------

# compute five number summary
# use genfromtxt to read file
import numpy as np
np.set_printoptions(precision=3,suppress=1,floatmode='maxprec')
tempvals = np.genfromtxt('tempstudy.csv',dtype=float,delimiter=',',\
                        usecols=(8), skip_header=True)
print('The number of values in tempvals is {}.'.format(len(tempvals)))
M = np.median(tempvals)
Q1 = np.quantile(tempvals, .25)
Q3 = np.quantile(tempvals, .75)
min = np.min(tempvals)
max = np.max(tempvals)
R = max - min
IQR = Q3-Q1
print('Minimum = {:.2f}'.format(min))
print('Q1      = {:.2f}'.format(Q1))
print('Median  = {:.2f}'.format(M))
print('Q3      = {:.2f}'.format(Q3))
print('Maximum = {:.2f}'.format(max))
print('Range   = {:.2f}'.format(R))
print('IQR     = {:.2f}'.format(IQR))
# compute the mean and standard deviation
xbar = np.mean(tempvals)
print('Mean    = {:.2f}'.format(xbar))
sd = np.std(tempvals,ddof=1)
```

```
print('Std.Dev.= {:.2f}'.format(sd))
# create histograms
plt.hist(tempvals, density=1,edgecolor="black")
plt.xlabel('Temperature')
plt.ylabel('Relative Frequency')
plt.show()


# ==========================================================

# compute Reimann sums
import numpy as np
import matplotlib.pyplot as plt
np.set_printoptions(precision=3,suppress=1,floatmode='maxprec')
# define the function
def stdnorm(x):
    y = 1 / np.sqrt(2 * np.pi) * np.exp(-x ** 2 / 2)
    return y

# get the number of rectangles
n = int(input('Enter the number of rectangles: '))
# limits of integration
a = -2.3
b = 1.2

# determine Delta x
dx = (b-a)/n
# create the partiion of x values
x = np.arange(a,b+dx,dx)

# get the y values (heights of the rectangles
y = stdnorm(x)

# Compute the areas of each rectangle
A = y[1:]*dx
# Sum the areas
R = np.sum(A)
print('Riemann Sum is:',R)
# ---------------------------------------------------------

import numpy as np
import matplotlib.pyplot as plt
np.set_printoptions(precision=3,suppress=1,floatmode='maxprec')
```

```python
# use the trapezoidal rule to approximate the integral of f(x) from a to b
# we call the the function traprule (trapezoidal rule)
def traprule(f,a,b,n):
    dx = (b-a)/n
    x = np.arange(a,b+dx,dx)
    y = f(x)
    # get the terms in the parentheses
    # multiply all but the first and last y-values by 2
    y[1:n] = 2*y[1:n]
    # now multiply all terms by (delta x)/2
    y = (dx/2)*y
    # sum the the areas
    T = np.sum(y)
    # return the value (T is the approximation to the integral)
    return T

# integrate the standard normal density function from -2.3 to 1.2 using
# 50 subintervals.
IntegralVal = traprule(stdnorm,-2.3,1.2,50)
print('The approximate value of the integral is',IntegralVal)

# ===========================================================

# confidence intervals with sigma known
import scipy.stats as stats
import numpy as np
c = stats.norm.ppf(q=0.025,loc=0,scale=1.0)
d = stats.norm.ppf(q=0.975,loc=0,scale=1.0)
print('c =',c)
print('d =',d)
xbar = 67.2
sig = 1.75
a = xbar - d*sig/np.sqrt(100)
b = xbar - c*sig/np.sqrt(100)
print('a =',a)
print('b =',b)
# -----------------------------------------------------------

# confidence interval using t distribution
import numpy as np
import scipy.stats as stats
```

```python
x = np.array([65.654, 67.263, 67.186, 64.808, 66.137, 67.487, 67.214, 72.155,
  69.201, 68.274, 67.610, 70.088, 66.167, 68.535, 66.216, 67.382, 66.867, 68.633,
  65.349, 69.423, 67.729, 67.250, 65.304, 68.566, 62.739, 65.567, 69.029, 67.769,
  62.608, 64.695, 66.873, 64.753, 70.209, 65.162, 66.258, 69.359, 69.038, 68.135,
  66.837, 67.007, 69.321, 67.853, 69.662, 65.779, 65.295, 66.136, 69.085, 69.504,
  67.754, 65.131, 66.470, 67.661, 68.761, 65.610, 67.970, 69.646, 69.795, 64.861,
  66.320, 67.531, 65.426, 66.926, 70.485, 67.880, 66.498, 68.265, 65.429, 68.368,
  66.464, 67.190, 70.934, 68.399, 68.986, 68.162, 65.521, 66.383, 66.250, 63.739,
  67.099, 63.716, 66.573, 62.929, 67.399, 66.959, 66.416, 68.436, 71.919, 66.320,
  67.314, 66.979, 67.733, 66.684, 67.074, 67.174, 68.305, 65.056, 67.582, 67.737,
  64.178, 70.572])
n = len(x)
xbar = np.mean(x)
print('xbar =',xbar)
s = np.std(x,ddof=1)
print('s =',s)
c = stats.t.ppf(0.025,n-1)
d = stats.t.ppf(0.975,n-1)
print('c=',c)
print('d=',d)
a = xbar - d*s/np.sqrt(n)
b = xbar - c*s/np.sqrt(n)
print('a =',a)
print('b =',b)
# ---------------------------------------------------------

# confidence interval using Python built-in methods
import numpy as np
import scipy.stats as stats

x = np.array([65.654, 67.263, 67.186, 64.808, 66.137, 67.487, 67.214, 72.155,
  69.201, 68.274, 67.610, 70.088, 66.167, 68.535, 66.216, 67.382, 66.867, 68.633,
  65.349, 69.423, 67.729, 67.250, 65.304, 68.566, 62.739, 65.567, 69.029, 67.769,
  62.608, 64.695, 66.873, 64.753, 70.209, 65.162, 66.258, 69.359, 69.038, 68.135,
  66.837, 67.007, 69.321, 67.853, 69.662, 65.779, 65.295, 66.136, 69.085, 69.504,
  67.754, 65.131, 66.470, 67.661, 68.761, 65.610, 67.970, 69.646, 69.795, 64.861,
  66.320, 67.531, 65.426, 66.926, 70.485, 67.880, 66.498, 68.265, 65.429, 68.368,
  66.464, 67.190, 70.934, 68.399, 68.986, 68.162, 65.521, 66.383, 66.250, 63.739,
  67.099, 63.716, 66.573, 62.929, 67.399, 66.959, 66.416, 68.436, 71.919, 66.320,
  67.314, 66.979, 67.733, 66.684, 67.074, 67.174, 68.305, 65.056, 67.582, 67.737,
  64.178, 70.572])
```

```
n = len(x)
xbar = np.mean(x)
s = np.std(x,ddof=1)
print('sample mean =',xbar)
print('sample standard deviation =',np.std(x,ddof=1))
print('standard dev of sample mean =',s/np.sqrt(n))
a,b = stats.t.interval(alpha=0.95, df=n-1, loc=xbar, scale=s/np.sqrt(n))
print('left end of interval =',a)
print('right end of interval =',b)
# =========================================================

# hypothesis tests
import numpy as np
import matplotlib.pyplot as plt
np.set_printoptions(precision=3,suppress=1,floatmode='maxprec')

x=np.array([2.46,2.2,2.09,2.84,2.82,2.19,2.76,2.72,2.98,2.22,
2.74,2.28,2.47,2.13,2.81,2.98,2.01,2.67,2.34,2.62])
xbar = np.mean(x)
s = np.std(x,ddof=1)
print('Sample mean of GPAs:        {:.3f}'.format(xbar))
print('Sample standard deviation: {:.3f}'.format(s))
n = len(x)  #number of oberservations in the list
# get the cumulative distribution for xbar
cdf = stats.t.cdf(xbar,df=n-1,loc=2.35,scale=s/np.sqrt(n))
print('cdf(2.516) = {:.4f}'.format(cdf))
# ----------------------------------------------------------

# hypothesis test using critical value
import numpy as np
import matplotlib.pyplot as plt
np.set_printoptions(precision=3,suppress=1,floatmode='maxprec')
x=np.array([2.46,2.2,2.09,2.84,2.82,2.19,2.76,2.72,2.98,2.22,
2.74,2.28,2.47,2.13,2.81,2.98,2.01,2.67,2.34,2.62])
xbar = np.mean(x)
s = np.std(x,ddof=1)
print('Sample mean of GPAs:        {:.3f}'.format(xbar))
print('Sample standard deviation: {:.3f}'.format(s))
n = len(x)  #number of oberservations in the list
GPA_crit = stats.t.ppf(0.95,df=n-1,loc=2.35,scale=s/np.sqrt(n))
print('The critical value for GPA is {:.3f}.'.format(GPA_crit))
```

```
# =============================================================

# comparing two means
import numpy as np
np.set_printoptions(precision=3,suppress=1,floatmode='maxprec')
import scipy.stats as stats

A = [2.61,-4.29,-2.5,2.34,-4.31,-4.37,2.01,0.16,1.04,3.13]
B = [0.62,-4.03,-5.32,-6.92,-0.84,-4.88,-7.83,-4.84,2.07,-4.88]
nA = len(A)
nB = len(B)
xbarA = np.mean(A)
sdA = np.std(A,ddof=1)
xbarB = np.mean(B)
sdB = np.std(B,ddof=1)
xdiff = xbarA - xbarB
pooledVar = ((nA-1)*sdA**2+(nB-1)*sdB**2)/(nA+nB-2)
sp = np.sqrt(pooledVar)
dofp = nA+nB-2
teststat = xdiff/(sp*np.sqrt(1/nA+1/nB))
print('test statistic:',teststat)
# mu_A < mu_B (group A lost more weight than B)
pvalL = stats.t.cdf(teststat,dofp)
# mu_A > mu_B (group A lost less weight than B)
pvalG = 1-stats.t.cdf(teststat,dofp)
pval = 2*np.min([pvalL,pvalG])
print('The p value is {:.3f}.'.format(pval))
# -----------------------------------------------------------

# hypothesis test using built-in methods
import numpy as np
np.set_printoptions(precision=3,suppress=1,floatmode='maxprec')
import scipy.stats as stats

A = [2.61,-4.29,-2.5,2.34,-4.31,-4.37,2.01,0.16,1.04,3.13]
B = [0.62,-4.03,-5.32,-6.92,-0.84,-4.88,-7.83,-4.84,2.07,-4.88]
print(stats.ttest_ind(a=A, b=B, equal_var=True))
# -----------------------------------------------------------

# paired t test
import numpy as np
np.set_printoptions(precision=3,suppress=1,floatmode='maxprec')
```

```
import scipy.stats as stats

A = [152.61, 145.71, 147.5,  152.34, 145.69,\
     145.63, 152.01, 150.16, 151.04, 153.13]
B = [149.62, 144.97, 143.68, 142.08, 148.16,\
     144.12, 141.17, 144.16, 151.07, 144.12]
xbarA = np.mean(A)
print('Pre-diet sample mean: {:.4f}'.format(xbarA))
xbarB = np.mean(B)
print('Post-diet sample mean: {:.4f}'.format(xbarB))
t,p = stats.ttest_rel(a=A, b=B)
print('t-stat = {:.4f}, p-value = {:.4f}'.format(t,p))
print('')
print('Assuming Ha: mu1 != mu2')
print('p value = {:.4f}'.format(p))
print('')
diffofmeans = xbarA - xbarB
if diffofmeans>0:
    print('Assuming Ha: mu1 > mu2')
    print('p value = {:.4f}'.format(0.5*p))
else:
    print('Assuming Ha: mu1 < mu2')
    print('p value = {:.4f}'.format(0.5*p))
# ========================================================

# comparing more than two means, one-way ANOVA
import numpy as np

# load the data from the file
studydata = np.genfromtxt('studydata.csv', delimiter=',',skip_header=1)

# determine the number of rows and columns in the data
# each column is a group
# this code assumes that there are the same number of observations in each
# group
m,n = np.shape(studydata)

# compute the sample variance s^2 for each group
vars = np.zeros(n)
for i in range(n):
    vars[i] = np.var(studydata[:, i], ddof=1)
```

```python
# compute MSE
SSE = np.sum((m-1)*vars)
print('SSE: {:.4f}'.format(SSE))
MSE = SSE/(m*n-n)
print('MSE (within groups): {:.4f}'.format(MSE))

# compute MST
# get the means for each group
means = np.zeros(n)
for i in range(n):
    means[i] = np.mean(studydata[:, i])

#get the overall mean
xbar = np.mean(studydata)

# compute MST
SST = np.sum(m*(means - xbar)**2)
print('SST: {:.4f}'.format(SST))
MST = SST/(n-1)
print('MST (between groups): {:.4f}'.format(MST))
# Compute the F statistic
F = MST/MSE
print('F statistic: {:.4f}'.format(F))

# compute the p-value
p = 1-stats.f.cdf(F,n-1,m*n-n)
print('p-value: {:.5f}'.format(p))
```

## A.6  Chapter 7 code

```python
# linear regression
import numpy as np
import matplotlib.pyplot as plt

#open the data file
pov_file = open('poverty.txt','r')
#see how many lines are in the file
count = len(pov_file.readlines())
pov_file.close()
#since the first line contains headers, there is one less actual
#lines of data
```

```
count = count-1

#set up some storage space for the data
x = np.zeros((count,1))
y = np.zeros((count,1))
#now we will reopen the file and read it line by line
#the first line of this file is a header
pov_file = open('poverty.txt','r')
headers = pov_file.readline()
#i printed the headers just in case i wanted to reference them
print(headers)
#now read the rest of the file and store the x's and the y's
for i in range(count):
    #get the next line and store it in l
    l = pov_file.readline()
    #split the line into separate fields (assumes space delimited)
    fields = l.split()
    #the second field (which will have an index of 1) is the poverty percent.
    #this is our x value
    x[i] = float(fields[1])
    #the third field holds the births that we want.  store them in y
    y[i] = float(fields[2])
#close the file
pov_file.close()

#our variables are m and b.  we need the matrix of coefficients
A = np.zeros((2,2))
#first row of coefficients
A[0,0] = np.sum(x*x)
A[0,1] = np.sum(x)
#second row of coefficients
A[1,0] = np.sum(x)
#the sum of 1 is equal the number of terms in the sum
A[1,1] = len(x)
#now we need the right hand side
B = np.zeros(2)
B[0] = np.sum(x*y)
B[1] = np.sum(y)
print('A=',A)
print('B=',B)
#now solve the system X = [m b]
X = np.linalg.solve(A,B)
```

```python
print('X=',X)
#plot the regression line
m = X[0]
b = X[1]
yhat = m*x+b
plt.plot(x,y,'.')
plt.plot(x,yhat)
plt.legend(['data','regression line'])
plt.show()
# compute correlation
SSR = np.sum((yhat-ybar)*(yhat-ybar))
SST = np.sum((y-ybar)*(y-ybar))
print('SSR=',SSR)
print('SST=',SST)
R2 = SSR/SST
print('R squared=',R2)
print('r=',np.sqrt(R2))
# ---------------------------------------------------------


# multiple regression
import numpy as np

#open the data file
pov_file = open('poverty.txt','r')
#see how many lines are in the file
count = len(pov_file.readlines())
pov_file.close()
#since the first line contains headers, there is one less actual
#lines of data
count = count-1

#set up some storage space for the data
A = np.ones((count,3))
y = np.zeros((count,1))
#now we will reopen the file and read it line by line
#the first line of this file is a header
pov_file = open('poverty.txt','r')
headers = pov_file.readline()
#i printed the headers just in case i wanted to reference them
print(headers)
#now read the rest of the file
```

```python
for i in range(count):
    #get the next line
    l = pov_file.readline()
    #split the line into separate fields
    fields = l.split()
    #the second field is the poverty percent
    A[i,1] = float(fields[1])
    #the fifth field is the crime rate
    A[i,2] = float(fields[4])
    #the third field holds the births that we want.  store them in y
    y[i] = float(fields[2])
#close the file
pov_file.close()
#multiply both sides by A_transpose
A_trans = A.transpose()
C = np.dot(A_trans,A)
#right hand side
RHS = np.dot(A_trans, y)

#now solve the system
X = np.linalg.solve(C,RHS)
print('X=',X)
print('yhat = {:.4f} + {:.4f}(poverty) + {:.4f}(crime)'.\
      format(X[0,0],X[1,0],X[2,0]))

ybar = np.average(y)

yhat = np.dot(A,X)

SSR = np.sum((yhat[:,0]-ybar)*(yhat[:,0]-ybar))
SST = np.sum((y-ybar)*(y-ybar))
print('SSR=',SSR)
print('SST=',SST)
R2 = SSR/SST
print('R squared=',R2)
print('r=',np.sqrt(R2))
# compute coefficient of determination
ybar = np.average(y)
yhat = np.dot(A,X)
SSR = np.sum((yhat[:,0]-ybar)*(yhat[:,0]-ybar))
SST = np.sum((y-ybar)*(y-ybar))
print('SSR =',SSR)
```

```python
print('SST =',SST)
R2 = SSR/SST
print('R squared =',R2)
# ----------------------------------------------------------

# multiple regression with built-in methods
import numpy as np
from sklearn.linear_model import LinearRegression

# load the poverty percent into column 1 of A and the crime rate in column 2
# we do not need a column of ones because the method will do that for us
A = np.genfromtxt('poverty.txt',dtype=float,usecols=(1,4), skip_header=True)
# load the birth rates into Y
Y = np.genfromtxt('poverty.txt',dtype=float,usecols=(2), skip_header=True)

# now the data matrix A and the actual y values Y are complete
# fit the regression line and store it.
# we are storing it in a variable named birthmodel of type LinearRegression
birthmodel = LinearRegression()
# find the parameters for the regression line
birthmodel.fit(A, Y)
# get the coefficient of determination (R-squared_
R2 = birthmodel.score(A, Y)
print('R squared =',R2)
# variables of type LinearRegression have components called
# coef_ and intercept_ that store the coefficients and intercept of
# the model.
coeff = birthmodel.coef_
intercept = birthmodel.intercept_
#coeff = np.append(coeff,reg.intercept_)
#print(coeff)
print('yhat = ({:.4f})poverty + ({:.4f})crime + {:.4f}'\
      .format(coeff[0],coeff[1],intercept))
# ==========================================================

# logistic regression
import numpy as np
import matplotlib.pyplot as plt

#load the required tools for logistic regression
from sklearn.linear_model import LogisticRegression
from sklearn import metrics
```

```python
#create space for the height values.  This must be a matrix, even
#if there is only one variable.
x = np.zeros((50,1))
x[:,0] = np.array([78.29, 82.28, 81.36, 80.27, 76.48, 78.33, 82.05, 80.71,
 76.02, 79.44, 70.87, 75.32, 79.46, 74.52,  77.80, 74.90, 78.97, 81.12,
 83.04, 78.59,  75.73, 77.59, 75.69,  78.72, 76.32,74.22, 71.20, 73.85,
 74.20, 74.35, 68.99, 69.45, 70.77, 76.87, 73.29, 70.61, 75.85, 73.80,
 71.01, 72.65, 76.80, 74.39,  67.63, 70.95,  69.17, 69.94, 71.79, 72.04,
 70.69, 68.40])

#create storage space for status values
#this initalizes with 1's in all the values
y = np.ones(50)
#the second 25 values in the data are not basketball players
#so the status values should be replaced with 0's
y[25:50] = np.zeros(25)

#establish a regression object
BasketballRegr = LogisticRegression()

#perform the regression to fit the model
BasketballRegr.fit(x, y)

#find the percentage of correct classifications
score = BasketballRegr.score(x, y)
print('Percentage of correct predictions:',score*100)

#calculate the predicted values
yhat = BasketballRegr.predict(x)

#print the confusion matrix
print('Confusion Matrix')
cm = metrics.confusion_matrix(y, yhat)
print(cm)
# get parameters of the model
coeff = BasketballRegr.coef_
c_1 = coeff[0]
c_0 = BasketballRegr.intercept_
print('c_0 = ',c_0)
print('c_1 = ',c_1)
# ==============================================================
```

```
# use imshow
import numpy as np
import matplotlib.pyplot as plt
#plot a sample image
D = np.matrix([[ 0., 0., 0., 2., 13., 0., 0., 0.],
  [ 0., 0., 0., 8., 15., 0., 0., 0.],
  [ 0., 0., 5., 16., 5., 2., 0., 0.],
  [ 0., 0., 15., 12., 1., 16., 4., 0.],
  [ 0., 4., 16., 2., 9., 16., 8., 0.],
  [ 0., 0., 10., 14., 16., 16., 4., 0.],
  [ 0., 0., 0., 0., 13., 8., 0., 0.],
  [ 0., 0., 0., 0., 13., 6., 0., 0.]])

plt.imshow(D,cmap = plt.cm.gray)
plt.show()
# ----------------------------------------------------------

# digit recognition logistic regression
from sklearn.model_selection import train_test_split
from sklearn.datasets import load_digits
from sklearn.linear_model import LogisticRegression
from sklearn import preprocessing
from sklearn import metrics

#get the data for all the images
digits = load_digits()

#scales the data to help with numeric computation
data_scaled = preprocessing.scale(digits.data)

# Print to show there are 1797 images (8 by 8 images for a dimensionality
# of 64)
print('Image Data Shape' , digits.data.shape)
# Print to show there are 1797 labels (integers from 0--9)
print('Label Data Shape', digits.target.shape)

x_train, x_test, y_train, y_test = train_test_split(data_scaled,\
                              digits.target, test_size=0.25, random_state=0)

# all parameters not specified are set to their defaults
logisticRegr = LogisticRegression(max_iter=100)
```

```
#logisticRegr = LogisticRegression()

#do the logisitic regression
logisticRegr.fit(x_train, y_train)
# Use score method to get accuracy of model
score = logisticRegr.score(x_test, y_test)
print('Percent correct =',score)
predictions = logisticRegr.predict(x_test)
print('Number of predictions =',len(predictions))
cm = metrics.confusion_matrix(y_test, predictions)
print(cm)
# ===========================================================

# neural network
print('importing packages')
import numpy as np
from sklearn.model_selection import train_test_split
import matplotlib.pyplot as plt
from sklearn.preprocessing import StandardScaler
from sklearn import metrics
from sklearn.neural_network import MLPClassifier
print('packages imported')
print('reading data')
winedata = np.genfromtxt('wine.csv', delimiter=',')
m,n = winedata.shape
print('winedata dimensions: ',m,n)
x = winedata[:,1:]
y = winedata[:,0]
x_train, x_test, y_train, y_test = \
    train_test_split(x, y, test_size=0.25,random_state=0)

print('x_train.shape',x_train.shape)
print('y_train.shape',y_train.shape)
print('x_test.shape',x_test.shape)
print('y_test.shape',y_test.shape)
# scale the data
scaler = StandardScaler()
# Fit only to the training data
scaler.fit(x_train)
# Now apply the transformations to the data:
x_train = scaler.transform(x_train)
x_test = scaler.transform(x_test)
```

```python
# fit the network
print('fit to neural net')
winenetwork = MLPClassifier(hidden_layer_sizes=(10),max_iter=1000)
winenetwork.fit(x_train,y_train)
# predict and score
predictions = winenetwork.predict(x_test)
proport_correct = winenetwork.score(x_test, y_test)
print('proportion of correct predictions',proport_correct)
# get the confusion matrix
cm = metrics.confusion_matrix(y_test, predictions)
print('Confusion Matrix:')
print(cm)
# classification report
print(metrics.classification_report(y_test, predictions))
# pretty confusion matrix
rowsums = np.sum(cm,0)
scaledcm = cm/rowsums
plt.imshow(scaledcm,cmap ='Blues',alpha=0.75)
plt.xticks(np.arange(0,3,1),['1','2','3'])
plt.yticks(np.arange(0,3,1),['1','2','3'])
for i in range(3):
    for j in range(3):
        plt.text(i-.1,j+.05,str(cm[i,j]))
plt.show()
```

# B Solutions

## Chapter 2. The basic operations in Python

**1.**
```
# 1a
print('Answer to Chapter 2, Number 1a')
print((4.1**2-4**2)/0.1)

# 1b
print('Answer to Chapter 2, Number 1b')
print((3+2)**3*(5-1)**4)
```

**Output**
```
Answer to Chapter 2, Number 1a
8.099999999999987
Answer to Chapter 2, Number 1b
32000
```

**2.**
```
# 2
print('Answer to Chapter 2, Number 2')
print('The radius is {:.2f} and the volume is {:.3f}.'\
      .format(4.23, 4/3*3.14*4.23**3))
```

**Output**
```
Answer to Chapter 2, Number 2
The radius is 4.23 and the volume is 316.876.
```

## Chapter 3. Functions

**1.**
```
# 1.

#print a table of trig values
import numpy
print("angle |{:>5} |{:>5} |{:>5} |{:>5} |{:>5} |{:>5}\
 |{:>5} |{:>5} |{:>5} |".format(\
    '0','\u03C0/6','\u03C0/4','\u03C0/3',\
    '\u03C0/2','2\u03C0/3','3\u03C0/4','5\u03C0/6','\u03C0'))
print('----------------------------------------------------\
```

https://doi.org/10.1515/9783110776645-009

```
-----------------')
print("cos(x)|{:.4f}|{:.4f}|{:.4f}|{:.4f}|{:.4f}|{:.3f}|{:.3f}|\
{:.3f}|{:.3f}|".format(numpy.cos(0),numpy.cos(numpy.pi/6),\
    numpy.cos(numpy.pi/4),numpy.cos(numpy.pi/3),\
    numpy.cos(numpy.pi/2),numpy.cos(2*numpy.pi/3),\
    numpy.cos(3*numpy.pi/4),numpy.cos(5*numpy.pi/6),\
    numpy.cos(numpy.pi)))
print("sin(x)|{:.4f}|{:.4f}|{:.4f}|{:.4f}|{:.4f}|{:.4f}\
|{:.4f}|{:.4f}|{:.4f}|".format(numpy.sin(0),\
    numpy.sin(numpy.pi/6),numpy.sin(numpy.pi/4),\
    numpy.sin(numpy.pi/3),numpy.sin(numpy.pi/2),\
    numpy.sin(2*numpy.pi/3),numpy.sin(3*numpy.pi/4),\
    numpy.sin(5*numpy.pi/63), numpy.sin(numpy.pi)))
print("tan(x)|{:.4f}|{:.4f}|{:.4f}|{:.4f}|undef |{:.3f}\
|{:.3f}|{:.3f}|{:.3f}|".format(numpy.tan(0),\
    numpy.tan(numpy.pi/6),numpy.tan(numpy.pi/4),\
    numpy.tan(numpy.pi/3),\
    numpy.tan(2*numpy.pi/3),numpy.tan(3*numpy.pi/4),\
    numpy.tan(5*numpy.pi/6), numpy.tan(numpy.pi)))
```

**Output**

```
angle |   0  |  π/6 |  π/4 |  π/3 |  π/2 | 2π/3| 3π/4 | 5π/6 |    π |

-----------------------------------------------------------------------
cos(x)|1.0000|0.8660|0.7071|0.5000|0.0000|-0.500|-0.707|-0.866|-1.000|
sin(x)|0.0000|0.5000|0.7071|0.8660|1.0000|0.8660|0.7071|0.2468|0.0000|
tan(x)|0.0000|0.5774|1.0000|1.7321|undef |-1.732|-1.000|-0.577|-0.000|

Process finished with exit code 0
```

**2.**

```
# 2.

import numpy
width = 2
length = numpy.sqrt(5)
area = width*length
print('The area of a box with width {:.2f} and length {:.2f}\
 is {:.2f}.'.format(width, length, area))
```

**Output**

```
The area of a box with width 2.00 and length 2.24 is 4.47.
```

```
Process finished with exit code 0
```

**3.**
```
# 3.

firstname = 'Albert'
lastname = 'Einstein'
fullname = lastname+', '+firstname
print('firstname:',firstname)
print('lastname:',lastname)
print('fullname:',fullname)
```

**Output**
```
firstname: Albert
lastname: Einstein
fullname: Einstein, Albert

Process finished with exit code 0
```

**4.**
```
# 4.

basetext = 'Force is equal to the product of mass and acceleration.'
m = basetext.find('mass')
print(m)
print(basetext[m:m+4])
n = basetext.find('product')
print(basetext[n:])
```

**Output**
```
33
mass
product of mass and acceleration.

Process finished with exit code 0
```

**5.**
```
# 5. a-c

def height(t):
    y = -16*t**2+3*t +100
```

```
    return y

y2 = height(2)
print('The value of height at t=2 is {:.3f}.'.format(y2))
```

**Output**

```
The value of height at t=2 is 42.000.

Process finished with exit code 0
```

**6.**
```
# 6.

def bmi(height,weight):
    # height in m and weight in kg
    b = weight/height**2
    return b

h = 1.7
w = 68
b = bmi(h,w)
print('A person who is {:.1f} m tall and weighs {} kg has a bmi\
 of {:.2f}.'.format(h,w,b))
```

**Output**

```
A person who is 1.7 m tall and weighs 68 kg has a bmi of 23.53.

Process finished with exit code 0
```

**7.**
```
# 7.

def bmi(height,weight):
    # height in m and weight in kg
    b = weight/height**2
    return b

h = input('Enter height in inches: ')
w = input('Enter weight in pounds: ')
#convert to numbers
h = float(h)
w = float(w)
```

```
#convert height and weight
h_m = h*.0254
w_p = w*.4536
b = bmi(h_m,w_p)
print('{} inches is {:.2f} meters'.format(h,h_m))
print('{} pounds is {:.2f} kg'.format(w,w_p))

print('A person who is {:.1f} m tall and weighs {} kg has a bmi\
 of {:.2f}.'.format(h_m,w_p,b))
```

**Output**

```
Enter height in inches: 67
Enter weight in pounds: 150
67.0 inches is 1.70 meters
150.0 pounds is 68.04 kg
A person who is 1.7 m tall and weighs 68.04 kg has a bmi of 23.49.

Process finished with exit code 0
```

**8.**
```
# 8.

import numpy
def area(r):
    a = numpy.pi*r**2
    return a

r = input('Enter the radius: ')
r = float(r)
a = area(r)
print('The area of the circle with radius {} is {:.4f}'.format(r,a))
```

**Output**

```
Enter the radius: 4
The area of the circle with radius 4.0 is 50.2655

Process finished with exit code 0
```

**9.**
```
# 9.

import numpy as np
```

```
import matplotlib.pyplot as plt
x = np.arange(0,2*np.pi+.1,np.pi/4)
y = np.cos(x)
plt.plot(x,y,'go:')
plt.grid()
plt.xlabel('x')
plt.ylabel('cos(x)')
plt.show()
```

**Output**

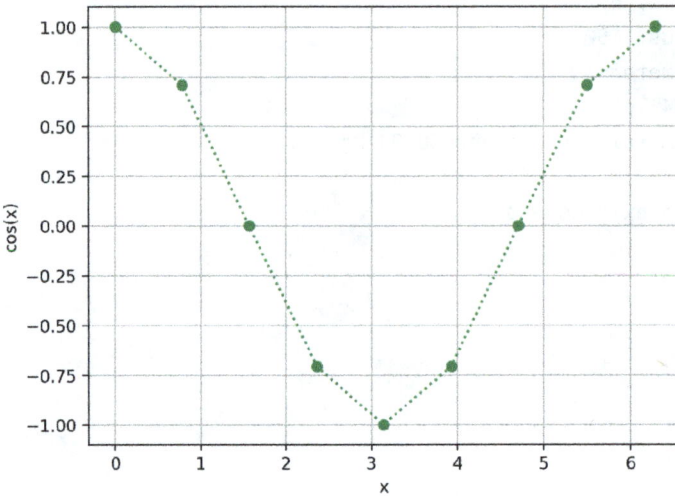

**10.**

```
# 10.
```

```
import numpy as np
import matplotlib.pyplot as plt
x = np.arange(0,2*np.pi+.1,np.pi/4)
y = np.cos(x)
y2 = np.sin(x)
plt.plot(x,y,'go:')
plt.plot(x,y2,'k')
plt.grid()
plt.xlabel('x')
plt.legend(['cos(x)','sin(x)'])
plt.show()
```

**Output**

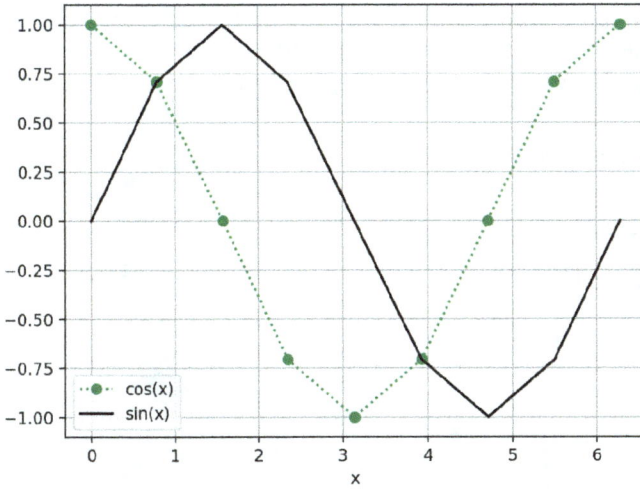

**11.**

```
# 11.

# a.
import numpy as np
import matplotlib.pyplot as plt

def f(x):
    y = -16*x**2+10*x+100
    return y

#use quadratic formula to solve -16t^2+10t+100
t1 = (-10+np.sqrt(10**2-4*(-16)*100))/(2*(-16))
t2 = (-10-np.sqrt(10**2-4*(-16)*100))/(2*(-16))
print('t1 =',t1, 'f(t1) =',f(t1))
print('t2 =',t2,'f(t2) =',f(t2))

# b.

x = np.arange(0,t2+t2/50,t2/50)
y = f(x)
plt.plot(x,y)
plt.grid()
plt.xlabel('time')
plt.ylabel('height')
```

```
plt.show()
```

**Output**

```
t1 = -2.2069555463432966 f(t1) = 1.4210854715202004e-14
t2 = 2.8319555463432966 f(t2) = 1.4210854715202004e-14


Process finished with exit code 0
```

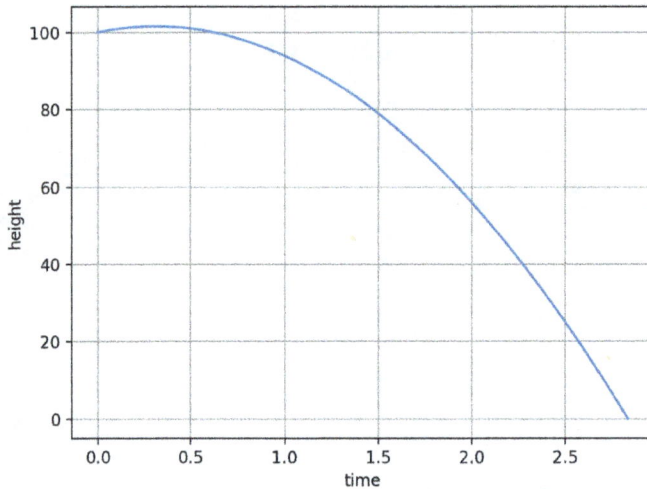

**12.**
#12.

```
import numpy as np
import matplotlib.pyplot as plt
def P(t):
    if 0<=t<3:
        y = np.exp(t)
    elif t>=0:
        y = (52.5614*t+3)/(t+5)
    else:
        print('invalid t value')
    return y
vP = np.vectorize(P)

dx = .01
x = np.arange(0,3,dx)
y = vP(x)
```

```
plt.plot(x,y)
x = np.arange(3,20+dx,dx)
y = vP(x)
plt.plot(x,y)
plt.show()
```

**b.** It does appear to be continuous

**b.** We would need the one-sided derivatives to be equal at $t = 3$

**Output**

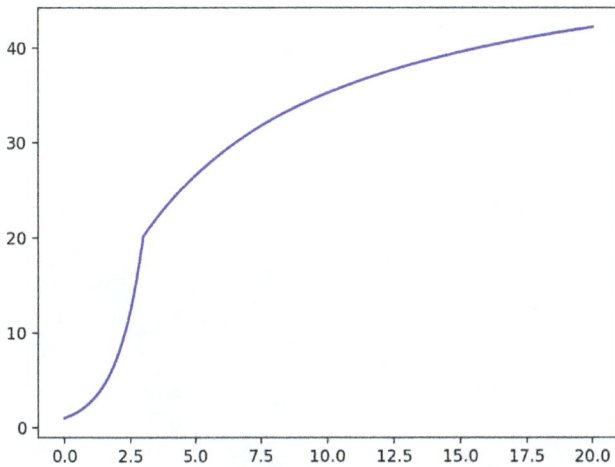

**13.**

```
# 13.

import numpy as np
import matplotlib.pyplot as plt

def f(x):
    if x<-1:
        y = x**2
    elif -1<=x<=1:
        y = x
    else:
        y = np.sin(x)
    return y
vf = np.vectorize(f)
```

```
dx = .01
x1 = np.arange(-2,-1,dx)
y1 = vf(x1)
plt.plot(x1,y1)
plt.plot(-1,1,'b.',fillstyle='none',markersize=11)
x2 = np.arange(-1,1,dx)
y2 = vf(x2)
plt.plot(x2,y2)
plt.plot(-1,-1,'b.',markersize=11)
plt.plot(1,1+dx,'b.',markersize=11)
x3 = np.arange(1+dx,3+dx,dx)
y3 = vf(x3)
plt.plot(x3,y3)
plt.plot(1,np.sin(1),'b.',fillstyle='none',markersize=11)
plt.xlabel('x')
plt.ylabel('y')
plt.grid()
plt.show()
```

**Output**

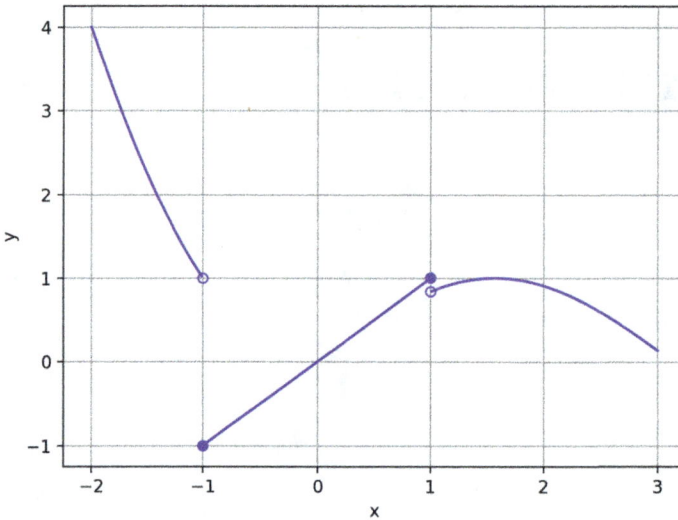

# Chapter 4. Matrices, vectors, and linear systems

**1.**

```
# 1.

import numpy as np

A = np.array([[1,-2,3],[2,1,4],[3,-1,-2]])
B = np.array([[0,4,2],[3,-1,-3]])
C = np.array([[-2,1],[0,-1],[1,3]])
D = np.array([[1,-3,0],[2,-2,2],[3,-1,1]])

print('A+D=')
print(A+D)
print('')
print('D-A=')
print(D-A)
print('')
print('3B=')
print(3*B)
print('')
print('A+B=')
#print(A+B)
print('Matrices must be same dimension')
print('')
print('Element-wise multiplication')
print('A*D=')
print(A*D)
print('')
print('A*C=')
#print(A*C)
print('Matrices must be same dimension')
print('')
print('Element-wise division')
print('A/D=')
print(A/D)
print('')
print('A/B=')
#print(A/B)
print('Matrices must be same dimension')
print('')
print('Standard Matrix Multiplication')
```

```
print('AD=')
print(np.dot(A,D))
print('We get a warning because one of the values of D is a zero')
print('')
print('AB=')
#print(np.dot(A,B))
print('Inner dimensions do not match')
print('')
print('BC=')
print(np.dot(B,C))
print('')
```

**Output**

```
A+D=
[[ 2 -5  3]
 [ 4 -1  6]
 [ 6 -2 -1]]

D-A=
[[ 0 -1 -3]
 [ 0 -3 -2]
 [ 0  0  3]]

3B=
[[ 0 12  6]
 [ 9 -3 -9]]

A+B=
Matrices must be same dimension

Element-wise multiplication
A*D=
[[ 1  6  0]
 [ 4 -2  8]
 [ 9  1 -2]]

A*C=
Matrices must be same dimension

Element-wise division
A/D=
[[ 1.         0.66666667         inf]
```

```
 [ 1.          -0.5         2.          ]
 [ 1.           1.         -2.         ]]

A/B=
Matrices must be same dimension

Standard Matrix Multiplication
AD=
[[  6  -2  -1]
 [ 16 -12   6]
 [ -5  -5  -4]]
We get a warning because one of the values of D is a zero

AB=
Inner dimensions do not match

BC=
[[ 2  2]
 [-9 -5]]

/Users/WillMiles/Desktop/_Courses/SciComp/SciCompBook/BookCode/
booksolutions.py:280: RuntimeWarning: divide by zero encountered in
true_divide
  print(A/D)

Process finished with exit code 0
```

**2.**

Append the following to the code from number 1.

```
np.set_printoptions(precision=3,suppress=1,floatmode='fixed')
A_inv = np.linalg.inv(A)
print('A_inv =')
print(A_inv)
print(np.dot(A,A_inv))
```

The additional output is given below.

**Output**
```
A_inv =
[[-0.044  0.156  0.244]
 [-0.356  0.244 -0.044]
 [ 0.111  0.111 -0.111]]
```

```
[[ 1.000  0.000 -0.000]
 [-0.000  1.000  0.000]
 [ 0.000 -0.000  1.000]]
```

```
Process finished with exit code 0
```

**3.**

a. yes
b. yes
c. no
d. no
e. no
f. no

**4.**

$$2x - y = 5$$
$$3x + 2y = 4$$
$$2(2x - y = 5)$$
$$3x + 2y = 4$$
$$4x - 2y = 10$$
$$3x + 2y = 4$$

Add the two equations to get

$$7x = 14 \Rightarrow x = 2.$$

Then back substitute

$$4(2) - 2y = 10 \Rightarrow -2y = 2 \Rightarrow y = -1.$$

**5.**

```
# 5.

import numpy as np
#set up the matrices
A = np.array([[1,-3,2,-1],[1,-4,5,2],[-1,3,-1,3],[3,2,-1,-1]])
B = np.array([6,13,-23,6])

#using inverses
A_inv = np.linalg.inv(A)
X = np.dot(A_inv,B)
print('x = {:.4f}, y = {:.4f}, z = {:.4f}, w = {:.4f}'.\
```

```
        format(X[0],X[1],X[2],X[3]))

# using the solve method
X = np.linalg.solve(A,B)
print('x = {:.4f}, y = {:.4f}, z = {:.4f}, w = {:.4f}'.\
        format(X[0],X[1],X[2],X[3]))
```

**Output**
```
x = -14.7059, y = 31.4706, z = 42.6471, w = -29.8235
x = -14.7059, y = 31.4706, z = 42.6471, w = -29.8235


Process finished with exit code 0
```

**6.**
```
# 6.


import numpy as np
# solve the system formed by AX = B
A = np.array([[1,-3,2],[2,-4,5],[-1,3,-2]])
B = np.array([6,13,-23])
# using the solve method
X = np.linalg.solve(A,B)
```

**Output**
```
Traceback (most recent call last):
  File "/Users/WillMiles/Desktop/_Courses/SciComp/SciCompBook/BookCode/
        booksolutions.py", line 343, in <module>
    X = np.linalg.solve(A,B)
  File "<__array_function__ internals>", line 5, in solve
  File "/Library/Frameworks/Python.framework/Versions/3.8/lib/python3.8/
        site-packages/numpy/linalg/linalg.py", line 393, in solve
    r = gufunc(a, b, signature=signature, extobj=extobj)
  File "/Library/Frameworks/Python.framework/Versions/3.8/lib/python3.8/
        site-packages/numpy/linalg/linalg.py",
            line 88, in _raise_linalgerror_singular
    raise LinAlgError("Singular matrix")
numpy.linalg.LinAlgError: Singular matrix


Process finished with exit code 1
```

When solving in Python an error message is returned indicating that the matrix is singular. Thus, there are either an infinite number of solutions or no solution at all. So, further analysis of the system is required. Multiplying the third equation by –1 gives

$$x - 3y - 2z = 23.$$

Thus, the left-hand sides of equations 1 and 3 are the same, but the right-hand sides are different. Therefore, there is no solution to this system.

## Chapter 5. Iteration

**1.**

```
# 1.


# a.
#      The variable iterationcount is added to the code
#      and incremented for each iteration.  then it is
#      returned in the return statement
#      Note the new call to the bisect function in the main
#      program.
import numpy as np
import matplotlib.pyplot as plt

def bisect(f,a,b,tol):
    #we will do the first iterate before our while loop starts so that we
    #have a value to test against the tolerance
    x = (a+b)/2
    iterationcount = 0
    while np.abs(f(x))>tol:
        #print('a={:.5f}  f(a)={:.5f},    b={:.5f}  f(b)={:.5f}, \
        #  x={:.5f}  f(x)={:.5f}'.format(a,f(a),b,f(b),x,f(x)))
        #now decide whether we replace a or b with x
        if f(a)*f(x) < 0:
            #root is between a and x so replace b
            b = x
        elif f(b)*f(x)<0:
            #root is between b and x so replace a
            a = x
        else:
            # in this case, f(x) must be 0 and we have found the root
            # so we will know the root value is x and we can end the loop
            break
```

```
        #recompute the approximation
        x = (a+b)/2
        iterationcount = iterationcount + 1
    return x,iterationcount

# b.

def f(x):
    y = x**3-100*np.cos(x)
    return y

# begin main program
x = np.linspace(-1,4,100)  # could also use np.arange
y = f(x)
plt.plot(x,y)
plt.grid()
plt.show()

# c.
x,itcount = bisect(f,-1,4,.001)
print('x = ',x)
print('Number of iterations needed:',itcount)
```

**Output**
```
x =  1.534637451171875
Number of iterations needed: 14

Process finished with exit code 0
```

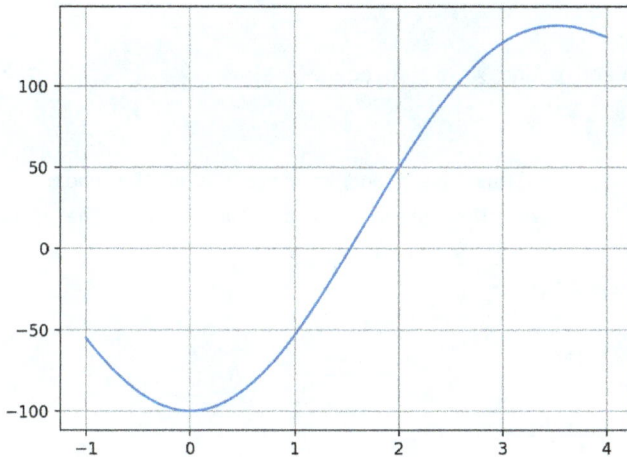

**2.**

```
# 2.

# a.
#      The variable iterationcount is added to the code
#      and incremented for each iteration.  then it is
#      returned in the return statement
#      Note the new call to the bisect function in the main
#      program.
import numpy as np
import matplotlib.pyplot as plt

def bisect(f,a,b,tol):
    #we will do the first iterate before our while loop starts so that we
    #have a value to test against the tolerance
    x = (a+b)/2
    iterationcount = 0
    # a variable to hold the absolute value of the difference
    # between two successive iterates.  We artifically initialize
    # it to a large value so that the loop is entered for the first
    # iterate
    diff = 1
    while diff > tol or iterationcount == 0:
        #print('a={:.5f}  f(a)={:.5f},   b={:.5f}  f(b)={:.5f}, \
        #   x={:.5f}  f(x)={:.5f}'.format(a,f(a),b,f(b),x,f(x)))
        #now decide whether we replace a or b with x
        if f(a)*f(x) < 0:
            #root is between a and x so replace b
            b = x
        elif f(b)*f(x)<0:
            #root is between b and x so replace a
            a = x
        else:
            # in this case, f(x) must be 0 and we have found the root
            # so we will know the root value is x and we can end the loop
            break
        # keep the previous iterate
        x_previous = x
        # compute the new iterate
        x = (a+b)/2
        # find the difference between them
```

```
        diff = np.abs(x-x_previous)
        iterationcount = iterationcount + 1
    return x,iterationcount
```

# b.

```
def f(x):
    y = x**3-100*np.cos(x)
    return y
```

```
x,itcount = bisect(f,-1,4,.001)
print('x = ',x)
print('Number of iterations needed:',itcount)
```

**Output**
```
Number of iterations needed: 12
```

```
Process finished with exit code 0
```

c. Comparing successive iterates will achieve more guaranteed accuracy.

**3.**
We require that

$$e^t = \frac{at+3}{t+5},$$

when $t = 3$. Thus, we need

$$e^3 = \frac{a3+3}{3+5}$$

$$e^3 - \frac{3a+3}{8} = 0.$$

So, we need the root of $f(x) = e^3 - \frac{3x+3}{8}$. This is a linear function, so we could solve this by hand. Begin by plotting to find an appropriate interval for the root. Trial and error shows there is a root between $x = 50$ and $x = 54$. Now, call the bisect function to calculate that the $a$ value is $a \approx 52.561462$.

# 3.

```
from bisectfun import *
```

```
def f(x):
    y = np.exp(3) - (3*x+3)/8
```

```
        return y

x = np.arange(50,60)
y = f(x)
plt.plot(x,y)
plt.grid()
plt.show()

x,itcount = bisect(f,50,54,.0001)
print('x = ',x)
print('Number of iterations needed:',itcount)

#check the answer
print('e^3 =',np.exp(3))
print('(3a+3)/(3+5) =',(x*3+3)/(8))
```

**Output**
```
x =  52.56146240234375
Number of iterations needed: 15
e^3 = 20.085536923187668
(3a+3)/(3+5) = 20.085548400878906

Process finished with exit code 0
```

**4.**

We have that $f(x) = e^x \sin(x) - \frac{x^2}{2} + 5$. Thus, $f'(x) = e^x \cos(x) + e^x \sin(x) - x$. So, we need to solve for the critical numbers, $f'(x) = 0$ and test the endpoints.

```
# 4.

from bisectfun import *

# original function
def f(x):
    y = np.exp(x) * np.sin(x)  - x**2/2 + 5
    return y

# derivative
def df(x):
    y = np.exp(x)*(np.sin(x)+np.cos(x)) - x
    return y
```

```
# plotting f' just to make sure there is a zero
dx = 0.1
x = np.arange(-1,3+dx,dx)
y = df(x)
plt.plot(x,y)
plt.grid()
plt.show()


x,itcount = bisect(df,-1,3,.0001)
print('x = ',x)
print('Number of iterations needed:',itcount)


# check critical numbers and endpoints in the original function
x_compare = [-1,3,x]
y_compare = [f(-1),f(3),f(x)]


# determine which index holds the max value
max_i = np.argmax(y_compare)
print('Max of {:.4f} occurs at x={:.4f}.'\
      .format(y_compare[max_i],x_compare[max_i]))
```

**Output**
```
x =   2.18121337890625
Number of iterations needed: 15
Max of 9.8787 occurs at x=2.1812.


Process finished with exit code 0
```

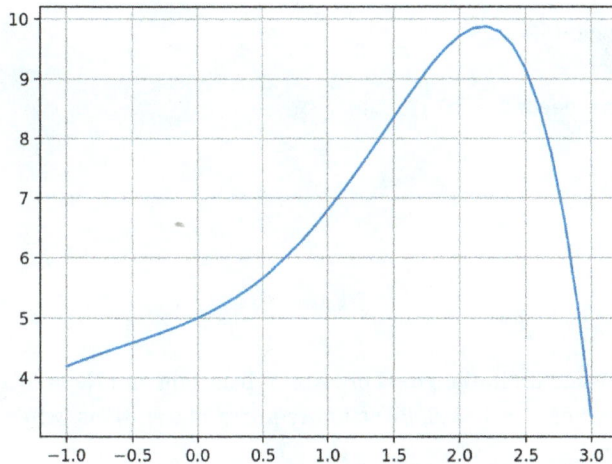

**5.**

Newton's method uses the following iterative function:

$$x_{n+1} = \frac{-f(x_n)}{f'(x_n)} + x_n.$$

```
# 5.

import numpy as np
# Newton's method
# f is the original function
# df is the derivative of f
# x0 is the initial guess for the root
def newtonroot(f,df,x0,tol):
    itcount = 0
    while np.abs(f(x0)) > tol:
        x1 = -f(x0)/df(x0)+x0
        itcount = itcount + 1
        x0 = x1
    return x0,itcount

def f(x):
    y = x**3-100*np.cos(x)
    return y

def df(x):
    y = 3*x**2+100*np.sin(x)
    return y

x,n = newtonroot(f,df,1,.00001)
print('x = ',x)
print('Number of iterations needed:',n)
```

**Output**
```
x =  1.5346454577941748
Number of iterations needed: 3


Process finished with exit code 0
```

For a tolerance of tol = 0.001 and an initial guess of x0 = 1, bisection and Newton's method return the same answer of $x$ = 1.5346. Bisection requires 14 iterations, while Newton's method requires only 3.

When the initial guess is $x_0 = -1$, then Newton's method converges to a root that is not in the desired interval.

**6.**

a.

```python
# Euler's method
import numpy as np
import matplotlib.pyplot as plt

# rhs is the right hand side function
# [a,b] is the interval of solution
# y0 is the initial value, y(a)
# dt is the step-size
def euler(rhs,a,b,y0,dt):
    #create a vector of t-values
    t = np.arange(a,b+dt,dt)
    n = len(t)
    #create space for the y-values
    y = np.zeros(n)
    #create a list of indices
    i = np.arange(1,n,1)
    #we know the inital value of y to be 1
    y[0] = y0
    for k in i:
        #compute the Euler approximation
        #use the right hand side function to get the slope of the
        #tangent line
        m = rhs(t[k-1],y[k-1])
        #get the next approximation
        y[k] = m*dt+y[k-1]
    return t,y

def rhs(t,y):
    m = t**2-np.sin(t)
    return m

def trusol(x):
    y = x**3/3 + np.cos(x) - 1
    return y

# Begin main program
t,y = euler(rhs,0,2*np.pi,0,.1)
```

```
#plot the solution
plt.plot(t,y,'b')
plt.autoscale(enable=True, axis='x', tight=True)
plt.xlabel('t')
plt.ylabel('y')
plt.grid()
plt.plot(t,trusol(t),'b--')
plt.legend(['Approximation','True Solution'])
plt.show()
```

**Output**

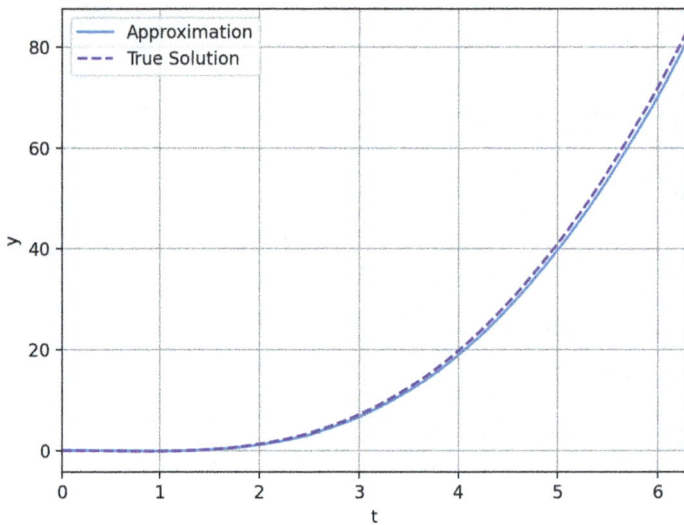

**7.**

Recall the general implicit Euler iteration

$$y_{n+1} = y'(t_{n+1}, y_{n+1})\Delta t + y_n.$$

So,

$$y_{n+1} = (\cos(t_{n+1}) + e^{-t_{n+1}} y_{n+1})\Delta t + y_n$$

$$y_{n+1} - e^{-t_{n+1}} y_{n+1}\Delta t = \cos(t_{n+1})\Delta t + y_n$$

$$y_{n+1}(1 - e^{-t_{n+1}}\Delta t) = \cos(t_{n+1})\Delta t + y_n$$

$$y_{n+1} = \frac{\cos(t_{n+1})\Delta t + y_n}{1 - e^{-t_{n+1}}\Delta t}$$

Then, one implementation is presented here:

```python
import numpy as np
import matplotlib.pyplot as plt

def rhs(t,y):
    r = np.cos(t)+ np.exp(-t)*y
    return r

a = 0
b = 2*np.pi
dt = 0.05
y0 = 0
t,y = euler(rhs,a,b,y0,dt)
plt.plot(t,y)

#implicit Euler
n = len(t)
y2 = np.zeros(n)
y2[0] = y0
for i in range(1,n,1):
    y2[i] = (np.cos(t[i])*dt+y2[i-1])/(1-np.exp(-t[i])*dt)
plt.plot(t,y2,'b--')
plt.legend(['Explicit Euler','Implicit Euler'])
plt.show()
```

**Output**

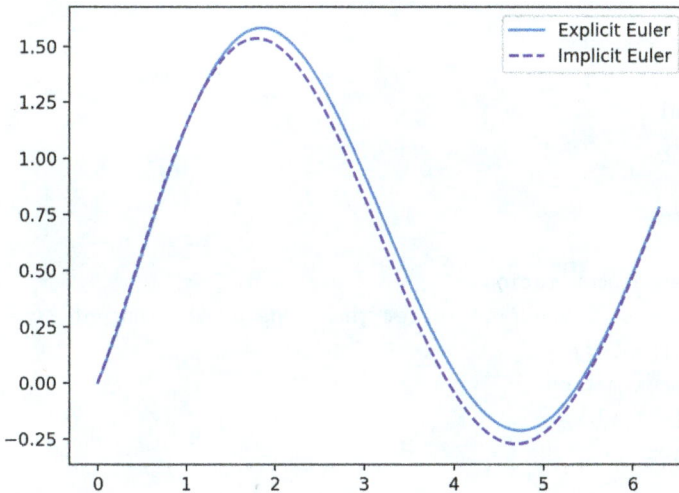

**8.**

```python
# Euler's method for systems
import numpy as np
import matplotlib.pyplot as plt

def rhs(t,yvec):
    a = 0.04
    b = 0.0005
    c = -0.1
    d = 0.0005
    dy = np.zeros(2)
    dy[0] = a*yvec[0]-b*yvec[0]*yvec[1]
    dy[1] = c*yvec[1]+d*yvec[0]*yvec[1]
    return dy

#initial t value
a = 0
#final t value
b = 365
#number of intervals
n = 50000
#delta t
dt = (b-a)/n
#create a vector of t-values
t = np.arange(a,b+dt/2,dt)

#create space for the y-values
y = np.zeros((n+1,2))
#create a list of indices
i = np.arange(1,n+1,1)
y[0,0] = 50
y[0,1] = 10
for k in i:
    #compute the Euler approximation
    #use the right hand side function to get the slope of the tangent line
    dy = rhs(t[k-1],y[k-1,:])
    #get the next approximation
    y[k,:] = dy*dt+y[k-1,:]
#plot the approximations
plt.plot(t,y[:,0])
plt.plot(t,y[:,1])
```

```
#plot true solution
plt.autoscale(enable=True, axis='x', tight=True)
plt.xlabel('t')
plt.grid()
plt.legend(['x(t)','y(t)'])
plt.show()
# ----------------------------------------------------------

# phase portrait
plt.figure()
plt.plot(y[:,0],y[:,1])
head = 1
tail = 0
w = 55
dx = y[head,0]-y[tail,0]
dy = y[head,1]-y[tail,1]
plt.arrow(y[head,0],y[head,1],dx,dy,width=.004)
numarrows = int((n-head)/w)
for i in range(4):
    head = head + w
    tail = tail + w
    dx = y[head,0]-y[tail,0]
    dy = y[head,1]-y[tail,1]
    plt.arrow(y[head,0],y[head,1],dx,dy,width=.004)
plt.xlabel('x(t)')
plt.ylabel('y(t)')
plt.title('Phase Portrait: IC = X(0) = 50, Y(0) = 10')
plt.grid()
plt.show()
```

**Output**

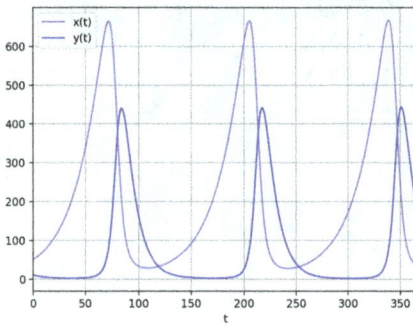
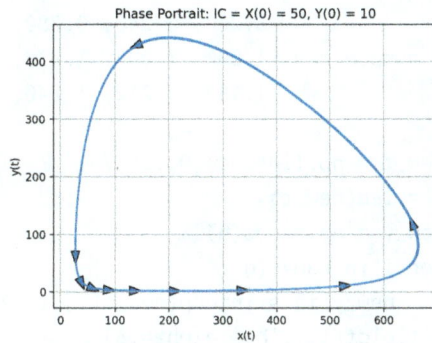

We can see that the populations are both periodic. The phase portrait shows that, while the predator population is low, the prey population grows easily. When the prey population rises above 600, the predator population begins to grow faster, and the prey population decreases more rapidly until the predator population is around 450, when the predator decreases quickly, and the cycle repeats.

9.

```python
import numpy as np
import matplotlib.pyplot as plt

def interp(x,t,y):
    n = len(t)
    startindex = 0
    # find the indices between which the new x value lies
    if x > np.max(t):
        print('outside of interpolation range')
        return np.nan
    elif x<np.min(t):
        print('outside of interpolation range')
        return np.nan
    while t[startindex]<x:
        startindex = startindex + 1
    startindex = startindex - 1
    endindex = startindex +1
    # slope for interpolation
    m = (y[endindex]-y[startindex])/(t[endindex]-t[startindex])
    # compute approximation using point slope form
    y_of_x = m*(x-t[startindex])+y[startindex]
    return y_of_x

t = np.array([0.000, 0.040, 0.080, 0.120, 0.160,\
              0.200, 0.240, 0.280, 0.320, 0.360])
y = np.array([1.000, 1.040, 1.078, 1.115, 1.150,\
              1.182, 1.212, 1.240, 1.266, 1.289])

new_t = np.linspace(0,.36,76)
n = len(new_t)
new_y = np.zeros(n)
for i in range(n):
    new_y[i] = interp(new_t[i],t,y)
plt.plot(t,y,'b-',alpha=.5)
```

```
plt.plot(new_t, new_y,'b-',alpha=.95)
plt.xlabel('t')
plt.ylabel('y')
plt.show()
```

**Output**

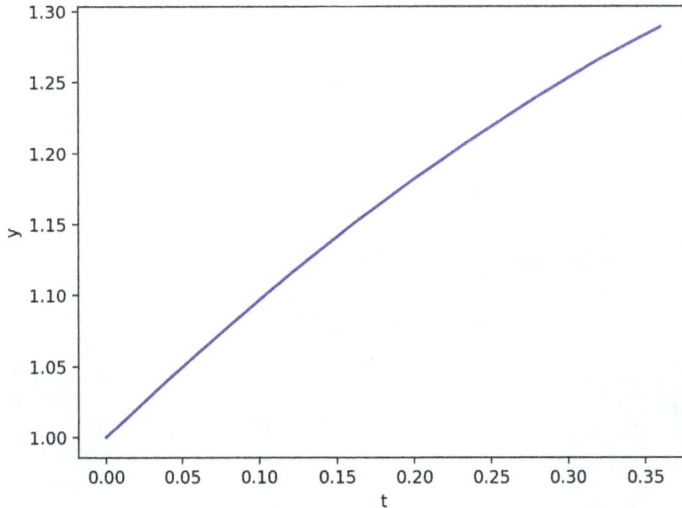

The graphs are indistinguishable.

# Chapter 6. Statistics

**1.**
**a.**

```
# file handling
# splitting strings and cleaning data
tumor_file = open('wdbc-ex.csv','r')
coded_file = open('wdbc-coded.csv','w')
# since the file is not very big, we can read the entire file
# into a list
print('loading data into list')
data = tumor_file.readlines()
tumor_file.close()
# let's see how many records there are
n = len(data)
```

```
print(n,'records in the list')
for i in range(n):
    # check the second field for an 'M' or a 'B'
    record = data[i]
    rec_vec = record.split(',')
    # get the number of fields in the record for later
    m = len(rec_vec)
    testfield = rec_vec[1]
    if testfield == 'M':
        # reassing the value to a one
        testfield = '1'
    elif testfield == 'B':
        # reassing the value to a one
        testfield = '0'
    # replace the M or B with the appropriate number
    rec_vec[1] = testfield
    # build the string to write to new file
    s = ''
    for j in range(m):
        if j < m-1:
            s = s+rec_vec[j]+','
        else:
            s = s+rec_vec[j]
    coded_file.write(s)
coded_file.close()
```

If the file were large so that it could not all be stored in a matrix, then the we would have to enter the file one line at a time. The changes in the code are minimal.

```
# file handling
# splitting strings and cleaning data
tumor_file = open('wdbc-ex.csv','r')
coded_file = open('wdbc-coded.csv','w')
# if the file is large, we must handle it line by line
record = tumor_file.readline()
while record != '':
    # check the second field for an 'M' or a 'B'
    rec_vec = record.split(',')
    # get the number of fields in the record for later
    m = len(rec_vec)
    testfield = rec_vec[1]
    if testfield == 'M':
```

```
            # reassing the value to a one
            testfield = '1'
        elif testfield == 'B':
            # reassing the value to a one
            testfield = '0'
        # replace the M or B with the appropriate number
        rec_vec[1] = testfield
        # build the string to write to new file
        s = ''
        for j in range(m):
            if j < m-1:
                s = s+rec_vec[j]+','
            else:
                s = s+rec_vec[j]
        coded_file.write(s)
        # get the next input record
        record = tumor_file.readline()
tumor_file.close()
coded_file.close()
```

**b**. We should use the file created in part (a) so that we do not undo the coding that was
done. The program should remove 15 records.

```
# remove records with missing data
tumor_file = open('wdbc-coded.csv','r')
coded_file = open('wdbc-coded2.csv','w')
print('loading data into list')
data = tumor_file.readlines()
tumor_file.close()
# let's see how many records there are
n = len(data)
print(n,'records in the list')
numdeleted = 0
for i in range(n):
    # check the second field for an 'M' or a 'B'
    record = data[i]
    rec_vec = record.split(',')
    # get the number of fields in the record for later
    m = len(rec_vec)
    deleteflag = 0
    for j in range(m):
        if rec_vec[j]=='':
```

```
            deleteflag = -1
            numdeleted = numdeleted+1
    if deleteflag == 0:
        #write the record
        coded_file.write(record)
print(numdeleted,'records deleted')
coded_file.close()
```

## Output
```
loading data into list
569 records in the list
loading data into list
569 records in the list
15 records deleted

Process finished with exit code 0
```

2.

There are a few challenges in this problem, and there are many approaches that would be acceptable. We first found that the years in the file were between 2006 and 2016. This allows one to set up the proper storage structure for the summary numbers. Then, one must create a formula to determine which row of the storage matrix is to be modified for each record read from the data file.

```
# computing averages of categories
import numpy as np
import matplotlib.pyplot as plt
#np.set_printoptions(precision=3,suppress=1,floatmode='fixed')
np.set_printoptions(suppress=1)


# open the temp study file for reading
temperature_file = open('tempstudy.csv','r')
# we know the first line of this file contains headers of the columns
record = temperature_file.readline()
# set up a matrix to hold what we need
# we need a row for each month and columns for
# total precip, total temperature, and number of observations
# the file includes the years 2006 - 2016.  so there will need
# to be 12 rows for each year, 12*11 = 132
# i'll store 2006 in rows 0-11, 2007 in rows 12-23, ...
tempsummary = np.zeros((132,4))
# read lines until you reach a blank line, then assume you are done
```

```
temprec = temperature_file.readline()
count = 0
while temprec != '':
    if count%1000000 == 0:
        print(count)
    # split the record into fields
    tempvec = temprec.split(',')
    # determine what years are in the file
    # get the year, month, precipitation, and temperature for
    # this observation
    year = int(tempvec[3])
    mth = int(tempvec[4])
    precip = float(tempvec[7])
    temp = float(tempvec[8])
    # determine the row of the summary matrix
    # (year-2005)*12 + mth-1
    row = (year-2006)*12 + mth-1
    tempsummary[row,0] = year
    tempsummary[row,1] = tempsummary[row,1] + precip
    tempsummary[row,2] = tempsummary[row,2] + temp
    tempsummary[row,3] = tempsummary[row,3] + 1
    temprec = temperature_file.readline()
    count = count + 1
temperature_file.close()
tempsummary[:,1] = np.round(tempsummary[:,1]/tempsummary[:,3],3)
tempsummary[:,2] = np.round(tempsummary[:,2]/tempsummary[:,3],3)
print(tempsummary)
# graph results
m = np.linspace(1,132,132)
plt.plot(m,tempsummary[:,1])
plt.xlabel('Month, 1=Jan, 2006')
plt.ylabel('precip')
plt.grid()
plt.figure()
plt.plot(m,tempsummary[:,2])
plt.xlabel('Month, 1=Jan, 2006')
plt.ylabel('temp')
plt.grid()
plt.show()
# write info to file
tempsum_file = open('tempsumm.csv','w')
for i in range(132):
```

```
      summrec = str(tempsummary[i,0])+','+str(tempsummary[i,1])+','\
              +str(tempsummary[i,2])+','+str(tempsummary[i,3])+'\n'
      tempsum_file.write(summrec)
tempsum_file.close()
```

**Output**

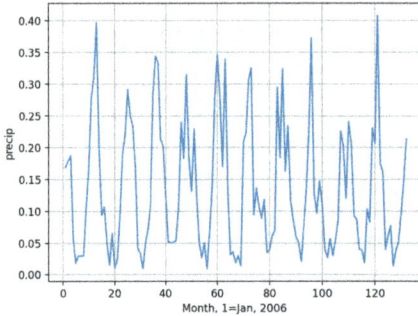
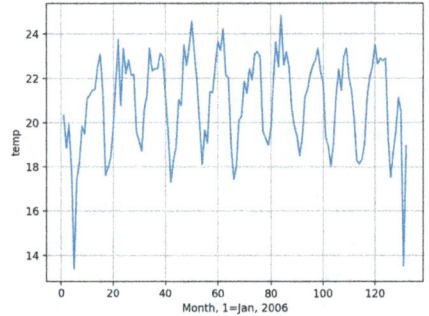

**3.**

```
import numpy as np
import matplotlib.pyplot as plt
#np.set_printoptions(precision=3,suppress=1,floatmode='fixed')
np.set_printoptions(suppress=1)


# read the file.  we have multiple ways to do this
# because the file is not large, genfromtxt is used here
tumor_data = np.genfromtxt('wdbc.csv',delimiter=',')
m,n = np.shape(tumor_data)
print('Data Matrix is {} x {}.'.format(m,n))
# radius is index 2
# texture is index 3
# smoothness is index 6
fivenumsummary = np.zeros((5,3))
datacols = [2,3,6]
# 3.b.
# generate the five number summaries, each column is a summary for
# a variable.
for i in range(3):
    col = datacols[i]
    fivenumsummary[0,i] = np.min(tumor_data[:,col])
    fivenumsummary[1,i] = np.quantile(tumor_data[:,col],.25)
```

```
        fivenumsummary[2,i] = np.median(tumor_data[:,col])
        fivenumsummary[3,i] = np.quantile(tumor_data[:,col],.75)
        fivenumsummary[4,i] = np.max(tumor_data[:,col])
# 3.c
plt.hist(tumor_data[:,2],density=1,bins=20,edgecolor = "black")
plt.xlabel('Radius')
plt.ylabel('Relative Frequency')

plt.figure()
plt.hist(tumor_data[:,3],density=1,bins=20,edgecolor = "black")
plt.xlabel('Texture')
plt.ylabel('Relative Frequency')

plt.figure()
plt.hist(tumor_data[:,6],density=1,bins=20,edgecolor = "black")
plt.xlabel('Smoothness')
plt.ylabel('Relative Frequency')
plt.show()

# 3.d.
# histograms are labeled and displayed above
# display the five-number summaries
variables = ['Radius','Texture','Smoothness']
for i in range(3):
    print('5-Number Summary for {}'.format(variables[i]))
    print('Minimum: {}'.format(fivenumsummary[0,i]))
    print('Q1      : {}'.format(fivenumsummary[1,i]))
    print('Median : {}'.format(fivenumsummary[2,i]))
    print('Q3      : {}'.format(fivenumsummary[3,i]))
    print('Maximum: {}'.format(fivenumsummary[4,i]))
    print('')
```

**Output**

```
Data Matrix is 569 x 32.
5-Number Summary for Radius
Minimum: 6.981
Q1      : 11.7
Median : 13.37
Q3      : 15.78
Maximum: 28.11


5-Number Summary for Texture
```

```
Minimum: 9.71
Q1     : 16.17
Median : 18.84
Q3     : 21.8
Maximum: 39.28

5-Number Summary for Smoothness
Minimum: 0.05263
Q1     : 0.08637
Median : 0.09587
Q3     : 0.1053
Maximum: 0.1634

Process finished with exit code 0
```

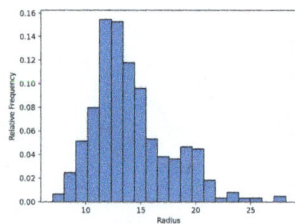

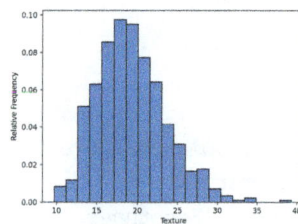

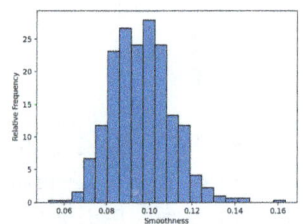

**4.**

```
# 4. a.

# compute Reimann sums
import numpy as np
import matplotlib.pyplot as plt
np.set_printoptions(precision=3,suppress=1,floatmode='maxprec')

def Rsum(fname,a,b,n):

    dx = (b-a)/n
    # create the partiion of x values
    x = np.arange(a,b+dx,dx)
    # get the y values (heights of the rectangles
    y = fname(x)
    # Compute the areas of each rectangle
    A = y[1:]*dx
    # Sum the areas
```

```
        R = np.sum(A)
        return R

# 4.b.
def f(x):
    y = np.exp(x)-x**2
    return y

R = Rsum(f,0,3,10)
print('Approximate Integral Value =',R)

# 4.c

R = Rsum(f,0,3,1000)
print('n=1000, Approximate Integral Value =',R)

R = Rsum(f,0,3,2000)
print('n=2000, Approximate Integral Value =',R)
```

**Output**
```
Approximate Integral Value = 11.69629473536423
n=1000, Approximate Integral Value = 10.100675042722996
n=2000, Approximate Integral Value = 10.093103529418098

Process finished with exit code 0
```

**5.**
```
# 5.

# a.
import numpy as np

def simprule(f,a,b,n):
    if n%2 != 0:
        print('n must be even')
        return np.nan
    # apply coefficients to even indices
    dx = (b-a)/n
    x = np.arange(a,b+dx,dx)
    # initialize the sum to the first term
    S = f(a)
    for i in range(1,n,1):
```

```
        if i%2 == 1:
            # x_1,x_3,x_5 ...
            S = S + 4*f(x[i])
        elif i%2 == 0:
            # x_2, x_4, ....
            S = S + 2*f(x[i])
    # last term
    S = S+f(b)
    S = (dx/3)*S
    return S

# b.


def f(x):
    y = np.sin(x) - 2/x
    return y

# F(x) = -cos(x) - 2*ln(x)
TrueValue = -np.cos(4) - 2*np.log(4) - (-np.cos(1)-2*np.log(1))
print('Actual Value of Integral =',TrueValue)

# c.

from rightsum import *

rightsum = Rsum(f,1,4,20)
print('Right Reimann Sum =:',rightsum)

# d.

from traprule import *

trapsum = traprule(f,1,4,20)
print('Trapezoidal Sum =:',trapsum)

# e.

simpsum = simprule(f,1,4,20)
print('Simpson Sum =:',simpsum)

# f.
print('Absolute Error:')
```

```
print('Right Sum:',np.abs(TrueValue-rightsum))
print('Trap  Sum:',np.abs(TrueValue-trapsum))
print('Simp  Sum:',np.abs(TrueValue-simpsum))

# we see that Simpson's rule is far better than the others

# g.
print('For 500 Subintervals')
print('Absolute Error:')
print('Right Sum:',np.abs(TrueValue-Rsum(f,1,4,500)))
print('Trap  Sum:',np.abs(TrueValue-traprule(f,1,4,500)))
print('Simp  Sum:',np.abs(TrueValue-simprule(f,1,4,500)))
```

**Output**
```
Actual Value of Integral = -1.5786427955080296
Right Reimann Sum =: -1.7909468156456947
Trapezoidal Sum =: -1.6839829487430966
Simpson Sum =: -1.578671394402969
Absolute Error:
Right Sum: 0.21230402013766514
Trap  Sum: 0.10534015323506707
Simp  Sum: 2.8598894939557695e-05
For 500 Subintervals
Absolute Error:
Right Sum: 0.00030402725876355063
Trap  Sum: 9.206818415874451e-06
Simp  Sum: 7.746070451730702e-11


Process finished with exit code 0
```

**6.**
```
# 6.

# confidence intervals with sigma known
import scipy.stats as stats
import numpy as np
c = stats.norm.ppf(q=0.05,loc=0,scale=1.0)
d = stats.norm.ppf(q=0.95,loc=0,scale=1.0)
xbar = 72
sig = 7
n = 20
a = xbar - d*sig/np.sqrt(n)
```

```
b = xbar - c*sig/np.sqrt(n)
print('{}% CI is ({:.4f}, {:.4f})'.format(90,a,b))
```

**Output**
```
90% CI is (69.4254, 74.5746)


Process finished with exit code 0
```

7.
```
# 7.

import numpy as np
import matplotlib.pyplot as plt
import scipy.stats as stats
np.set_printoptions(suppress=1)

# read the file.  we have multiple ways to do this
# because the file is not large, genfromtxt is used here
tumor_data = np.genfromtxt('wdbc.csv',delimiter=',')
m,n = np.shape(tumor_data)
print('Data Matrix is {} x {}.'.format(m,n))
# radius is index 2
xbar = np.mean(tumor_data[:,2])
print('xbar =',xbar)
sx = np.std(tumor_data[:,2],ddof=1)
print('sx =',sx)
a,b = stats.t.interval(alpha=0.90, df=n-1, loc=xbar, scale=sx/np.sqrt(m))
print('{}% CI is ({:.4f}, {:.4f})'.format(90,a,b))
```

**Output**
```
Data Matrix is 569 x 32.
xbar = 14.127291739894552
sx = 3.5240488262120775
90% CI is (13.8768, 14.3778)


Process finished with exit code 0
```

8.
```
# 8.

import numpy as np
import matplotlib.pyplot as plt
```

```python
import scipy.stats as stats
np.set_printoptions(precision=3,suppress=1,floatmode='maxprec')


data = np.array([62,67,45,59,54,53,60,53,47,46,\
                 59,47,74,66,58,69,50,68,62,58,\
                 51,56,66,58,58,68,56,75,64,46,\
                 58,67,59,55,64,68,50,50,68,39])
# hypothesis tests


xbar = np.mean(data)
s = np.std(data,ddof=1)
mu = 45.6
print('H0: mu = 45.6')
print('Ha: mu > 45.6')
print('Hypothesized mean       : {:.3f}'.format(mu))
print('Sample mean             : {:.3f}'.format(xbar))
print('Sample standard deviation: {:.3f}'.format(s))
n = len(data)  #number of oberservations in the list
print('Number of data values    : {}'.format(n))
# get the cumulative distribution for xbar
cdf = stats.t.cdf(xbar,df=n-1,loc=mu,scale=s/np.sqrt(n))
print('cdf({})               : {:.4f}'.format(xbar,cdf))
# greater than test
alpha = 0.01
pval = 1-cdf
print('pval                    : {:.6f}'.format(pval))
if pval <= alpha:
    print('Reject H0')
else:
    print('Do not reject H0')
# ----------------------------------------------------------
print('-----')
print('OR using critical value approach')
# hypothesis test using critical value
print('Sample mean             : {:.3f}'.format(xbar))
print('Sample standard deviation: {:.3f}'.format(s))
crit = stats.t.ppf(1-alpha,df=n-1,loc=mu,scale=s/np.sqrt(n))
print('The critical value      : {:.3f}.'.format(crit))
# greater than test
if xbar >= crit:
    print('Reject H0')
else:
```

```
    print('Do not reject H0')
```

**Output**

```
H0: mu = 45.6
Ha: mu > 45.6
Hypothesized mean        : 45.600
Sample mean              : 58.325
Sample standard deviation: 8.544
Number of data values    : 40
cdf(58.325)              : 1.0000
pval                     : 0.000000
Reject H0
-----
OR using critical value approach
Sample mean              : 58.325
Sample standard deviation: 8.544
The critical value       : 48.877.
Reject H0


Process finished with exit code 0
```

**9.**

```
# 9.


# comparing two means
import numpy as np
np.set_printoptions(precision=3,suppress=1,floatmode='maxprec')
import scipy.stats as stats


datafile = open('courserounds.csv','r')
# get header line
record = datafile.readline()
record = datafile.readline()
course2 = np.zeros(1)
course3 = np.zeros(1)
count = 0
while record !='':
    rec_vec = record.split(',')
    if count == 0:
        course2[0] = int(rec_vec[1])
        course3[0] = int(rec_vec[2])
    else:
```

```
        if rec_vec[1] != '':
            course2 = np.append(course2,int(rec_vec[1]))
    if rec_vec[2] != '':
        course3 = np.append(course3, int(rec_vec[2]))
    record = datafile.readline()
    count = count+1
#print(course1)
#print(course2)
n2 = len(course2)
n3 = len(course3)
xbar2 = np.mean(course2)
sd2 = np.std(course2,ddof=1)
xbar3 = np.mean(course3)
sd3 = np.std(course3,ddof=1)
xdiff = xbar2 - xbar3
pooledVar = ((n2-1)*sd2**2+(n3-1)*sd3**2)/(n2+n3-2)
sp = np.sqrt(pooledVar)
dofp = n2+n3-2
teststat = xdiff/(sp*np.sqrt(1/n2+1/n3))
print('test statistic:',teststat)
# mu_A < mu_B (group A lost more weight than B)
print('Ha: mu2 < mu3')
pvalL = stats.t.cdf(teststat,dofp)
print('pvalue =',pvalL)
print('')
# mu_A > mu_B (group A lost less weight than B)
print('Ha: mu2 > mu3')
pvalG = 1-stats.t.cdf(teststat,dofp)
print('pvalue =',pvalG)
print('')
print('Ha: mu2 != mu3')
pval = 2*np.min([pvalL,pvalG])
print('The p value is {:.3f}.'.format(pval))
print('')
print('OR, using built-in methods')
t,p=stats.ttest_ind(a=course2, b=course3, equal_var=True)
print('Test Statistic:',t)
print('Two-tailed pvalue:',p)
print('One-tailed pvalue:',p/2)

print('Since p={:.4f} > .05, there is not sufficient evidence to reject H0.'\
    .format(p/2))
```

**Output**
```
test statistic: 1.6094543626515967
Ha: mu2 < mu3
pvalue = 0.9454326485590657

Ha: mu2 > mu3
pvalue = 0.054567351440934275

Ha: mu2 != mu3
The p value is 0.109.

OR, using built-in methods
Test Statistic: 1.6094543626515967
Two-tailed pvalue: 0.10913470288186862
One-tailed pvalue: 0.05456735144093431
Since p=0.0546 > .05, there is not sufficient evidence to reject H0.

Process finished with exit code 0
```

**10.**
```python
# 10.

# paired t test
import numpy as np
np.set_printoptions(precision=3,suppress=1,floatmode='maxprec')
import scipy.stats as stats

weightdata = np.genfromtxt('weightdata.csv',delimiter=',',skip_header=1)
A = weightdata[:,0]
B = weightdata[:,1]
xbarA = np.mean(A)
print('Pre-diet sample mean: {:.4f}'.format(xbarA))
xbarB = np.mean(B)
print('Post-diet sample mean: {:.4f}'.format(xbarB))
t,p = stats.ttest_rel(a=A, b=B)
print('t-stat = {:.4f}, p-value = {:.4f}'.format(t,p))
print('')
print('Assuming Ha: mu1 != mu2')
print('p value = {:.4f}'.format(p))
print('')
diffofmeans = xbarA - xbarB
```

```
if diffofmeans>0:
    print('Assuming Ha: mu1 > mu2')
    print('p value = {:.4f}'.format(0.5*p))
else:
    print('Assuming Ha: mu1 < mu2')
    print('p value = {:.4f}'.format(0.5*p))

print('The appropriate test is Ha: muA > muB.')
print('So the pvalue is {:.4f}.'.format(p/2))
print('Since {:.4f} < .05, reject H0.'.format(p/2))
print('The diet appears to be effective.')
```

**Output**
```
Pre-diet sample mean: 198.3800
Post-diet sample mean: 195.9800
t-stat = 2.8640, p-value = 0.0051

Assuming Ha: mu1 != mu2
p value = 0.0051

Assuming Ha: mu1 > mu2
p value = 0.0026
The appropriate test is Ha: muA > muB.
So, the pvalue is 0.0026.
Since 0.0026 < .05, reject H0.
The diet appears to be effective.


Process finished with exit code 0
```

**11.**
```
# 11.


# comparing more than two means, one-way ANOVA
import numpy as np
import scipy.stats as stats

# load the data from the file
stockdata = np.genfromtxt('stockdata.csv', delimiter=',',skip_header=1)

# determine the number of rows and columns in the data
# each column is a group
# this code assumes that there are the same number of observations in
```

```
# each group
m,n = np.shape(stockdata)

# compute the sample variance s^2 for each group
vars = np.zeros(n)
for i in range(n):
    vars[i] = np.var(stockdata[:, i], ddof=1)

# compute MSE
SSE = np.sum((m-1)*vars)
print('SSE: {:.4f}'.format(SSE))
MSE = SSE/(m*n-n)
print('MSE (within groups): {:.4f}'.format(MSE))

# compute MST
# get the means for each group
means = np.zeros(n)
for i in range(n):
    means[i] = np.mean(stockdata[:, i])

#get the overall mean
xbar = np.mean(stockdata)

# compute MST
SST = np.sum(m*(means - xbar)**2)
print('SST: {:.4f}'.format(SST))
MST = SST/(n-1)
print('MST (between groups): {:.4f}'.format(MST))
# Compute the F statistic
F = MST/MSE
print('F statistic: {:.4f}'.format(F))

# compute the p-value
p = 1-stats.f.cdf(F,n-1,m*n-n)
print('p-value: {:.5f}'.format(p))

print('If p < \u03B1, we would reject the hypothesis that all the funds\
    have the same rate of return. ')
```

**Output**
```
SSE: 1514.1254
MSE (within groups): 7.7251
```

```
SST: 71.8462
MST (between groups): 23.9487
F statistic: 3.1001
p-value: 0.02787
```

If p < $\alpha$, we would reject the hypothesis that all the funds have the same rate of return.

```
Process finished with exit code 0
```

**12.**

```python
# a.
# comparing more than two means, one-way ANOVA
import numpy as np
import scipy.stats as stats

def OneWayANOVA(data):
    # determine the number of rows and columns in the data
    # each column is a group
    # this code assumes that there are the same number of observations
    # in each group
    m, n = np.shape(data)
    # compute the sample variance s^2 for each group
    vars = np.zeros(n)
    for i in range(n):
        vars[i] = np.var(data[:, i], ddof=1)

    # compute MSE
    SSE = np.sum((m-1)*vars)
    print('SSE: {:.4f}'.format(SSE))
    MSE = SSE/(m*n-n)
    print('MSE (within groups): {:.4f}'.format(MSE))

    # compute MST
    # get the means for each group
    means = np.zeros(n)
    for i in range(n):
        means[i] = np.mean(stockdata[:, i])

    #get the overall mean
    xbar = np.mean(data)
```

```
# compute MST
SST = np.sum(m*(means - xbar)**2)
print('SST: {:.4f}'.format(SST))
MST = SST/(n-1)
print('MST (between groups): {:.4f}'.format(MST))
# Compute the F statistic
F = MST/MSE
print('F statistic: {:.4f}'.format(F))

# compute the p-value
p = 1-stats.f.cdf(F,n-1,m*n-n)
print('p-value: {:.5f}'.format(p))
return F,p


# load the data from the file
stockdata = np.genfromtxt('stockdata.csv', delimiter=',',skip_header=1)

F,p = OneWayANOVA(stockdata)

print('If p < \u03B1, we would reject the hypothesis that all the funds\
    have the same rate of return. ')
```

**Output**
```
SSE: 1514.1254
MSE (within groups): 7.7251
SST: 71.8462
MST (between groups): 23.9487
F statistic: 3.1001
p-value: 0.02787
```

If $p < \alpha$, we would reject the hypothesis that all the funds have the same rate of return.

Process finished with exit code 0

**b.**

This is a challenging exercise. Using *args allows one to pass a variable number of arguments to a Python function. Then, we must iterate through the argument list to determine the number and composition of the arguments that have passed. One version of a modified ANOVA function is given in the following.

```python
# b.

# comparing more than two means, one-way ANOVA
import numpy as np
import scipy.stats as stats

def OneWayANOVA(*data):
    # determine the number of rows and columns in the data
    # each column is a group
    # this code assumes that there are the same number of observations
    # in each group
    n = len(data)
    # compute the sample variance s^2 for each group
    SSE = 0
    SST = 0
    #get the overall mean
    sum=0
    obs = 0
    for d in data:
        sum = sum+np.sum(d)
        obs = obs+len(d)
    xbar = sum/obs
    for d in data:
        m = len(d)
        variance = np.var(d, ddof=1)
        mean = np.mean(d)
        # compute MSE
        SSE = SSE + np.sum((m-1)*variance)
        # compute MST
        SST = SST+np.sum(m*(mean - xbar)**2)

    print('SSE: {:.4f}'.format(SSE))
    MSE = SSE / (obs - n)
    print('MSE (within groups): {:.4f}'.format(MSE))
    print('SST: {:.4f}'.format(SST))
    MST = SST / (n - 1)
    print('MST (between groups): {:.4f}'.format(MST))
    # Compute the F statistic
    F = MST/MSE
    print('F statistic: {:.4f}'.format(F))

    # compute the p-value
```

```
    p = 1-stats.f.cdf(F,n-1,obs-n)
    print('p-value: {:.5f}'.format(p))
    return F,p



# load the data from the file
stockdata = np.genfromtxt('stockdata.csv', delimiter=',',skip_header=1)
d1 = stockdata[:,0]
d2 = stockdata[:,1]
d3 = stockdata[:,2]
d4 = stockdata[:,3]
F,p = OneWayANOVA(d1,d2,d3,d4)

print('If p < \u03B1, we would reject the hypothesis that all the funds\
    have the same rate of return. ')
```

**Output**
```
SSE: 1514.1254
MSE (within groups): 7.7251
SST: 71.8462
MST (between groups): 23.9487
F statistic: 3.1001
p-value: 0.02787
```

If $p < \alpha$, we would reject the hypothesis that all the funds have the same rate of return.

```
Process finished with exit code 0
```

# Chapter 7. Regression

**1.**
```
# 1.
# multiple regression with built-in methods
import numpy as np
import matplotlib.pyplot as plt


# load the poverty percent into column 1 of A and the crime rate in
# column 2 we do not need a column of ones because the method will do that
# for us
x = np.genfromtxt('poverty.txt',dtype=float,usecols=(1), skip_header=True)
# load the birth rates into Y
```

```
y = np.genfromtxt('poverty.txt',dtype=float,usecols=(3), skip_header=True)
# linear regression

#our variables are m and b.  we need the matrix of coefficients
A = np.zeros((2,2))
#first row of coefficients
A[0,0] = np.sum(x*x)
A[0,1] = np.sum(x)
#second row of coefficients
A[1,0] = np.sum(x)
#the sum of 1 is equal the number of terms in the sum
A[1,1] = len(x)
#now we need the right hand side
B = np.zeros(2)
B[0] = np.sum(x*y)
B[1] = np.sum(y)
#now solve the system X = [m b]
X = np.linalg.solve(A,B)
print('predicted births = {:.4f}(poverty) + {:.4f}'.format(X[0],X[1]))
#plot the regression line
m = X[0]
b = X[1]
yhat = m*x+b
plt.plot(x,y,'.')
plt.plot(x,yhat,'b')
plt.legend(['data','regression line'])
plt.show()
# compute correlation
ybar = np.mean(y)
SSR = np.sum((yhat-ybar)*(yhat-ybar))
SST = np.sum((y-ybar)*(y-ybar))
print('SSR =',SSR)
print('SST =',SST)
R2 = SSR/SST
print('R squared = {:.4f}'.format(R2))
print('r = {:.4f}'.format(np.sqrt(R2)))
print('{:.4f}% of the variation is explained by the regression.'\
      .format(R2*100))
```

**Output**
```
predicted births = 2.8822(poverty) + 34.2124
SSR = 7598.568765118201
```

```
SST = 18003.60039215686
R squared = 0.4221
r = 0.6497
42.2058% of the variation is explained by the regression.


Process finished with exit code 0
```

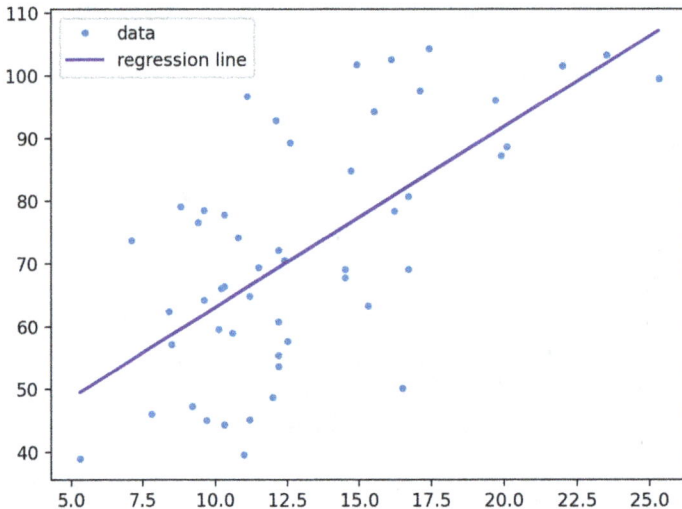

**2.**

```
# 2.


# multiple regression with built-in methods
import numpy as np
from sklearn.linear_model import LinearRegression


# load the explanatory variables into A
# we do not need a column of ones because the method will do that for us
A = np.genfromtxt('diabetes.txt',dtype=float,usecols=(0,1,2,3),\
                  skip_header=True)
# load the disease progression into Y
Y = np.genfromtxt('diabetes.txt',dtype=float,usecols=(10),\
                  skip_header=True)
# now the data matrix A and the actual y values Y are complete
# fit the regression line and store it.
# we are storing it in a variable named diseasemodel of type
```

```
# LinearRegression
diseasemodel = LinearRegression()
# find the parameters for the regression line
diseasemodel.fit(A, Y)
# get the coefficient of determination (R-squared_
R2 = diseasemodel.score(A, Y)
print('R squared =',R2)
print('r =',np.sqrt(R2))
# variables of type LinearRegression have components called
# coef_ and intercept_ that store the coefficients and intercept of
# the model.
coeff = diseasemodel.coef_
intercept = diseasemodel.intercept_
print('yhat = {:.4f}(x1) + {:.4f}(x2) + {:.4f}(x3) +\
{:.4f}(x4) + {:.4f}'.format(coeff[0],coeff[1],\
                                coeff[2],coeff[3],intercept))
print('')
print('R^2 and r indicate a moderate linear relationship.')
print('Coding the sex variable as 1 or 2 is likely inefficient.')
print('')
print('For woman (x2=2) with x1=45, x3=30, x4=112')
X = np.array([[45,2,30,112]])
yhat = diseasemodel.predict(X)
print('Disease Progression is:', np.round(yhat[0],3))
```

**Output**

```
R squared = 0.4002610119699945
r = 0.6326618464630173
yhat = 0.1353(x1) + -10.1590(x2) + 8.4843(x3) +1.4345(x4) + -199.0694

R^2 and r indicate a moderate linear relationship.
Coding the sex variable as 1 or 2 is likely inefficient.

For woman (x2=2) with x1=45, x3=30, x4=112
Disease Progression is: 201.899

Process finished with exit code 0
```

**3.**
```
# 3.

import numpy as np
```

```
from sklearn.linear_model import LinearRegression

# work must be done to separate the data for males and females.
# there are many ways to do this
# load the all data, including response values, into A
A = np.genfromtxt('diabetes.txt',dtype=float,usecols=(0,1,2,3,10),\
                  skip_header=True)
# now separate the men's data from the women's data
n = len(A)
combdata = np.zeros((n,3))
combdata[:,0] = A[:,0]
combdata[:,1:3] = A[:,2:4]
combresponse = A[:,4]
# count how many men are in the data
nummen = 0
for i in range(n):
    if A[i,1]==1:
        nummen = nummen +1
numwomen = n - nummen
mdata = np.zeros((nummen,3))
mresponse = np.zeros(nummen)
wdata = np.zeros((numwomen,3))
wresponse = np.zeros(numwomen)
mindex = 0
windex = 0
for i in range(n):
    if A[i,1]==1:
        mdata[mindex,0] = A[i,0]
        mdata[mindex, 1] = A[i,2]
        mdata[mindex, 2] = A[i,3]
        mresponse[mindex] = A[i,4]
        mindex = mindex + 1
    else:
        wdata[windex, 0] = A[i,0]
        wdata[windex, 1] = A[i,2]
        wdata[windex, 2] = A[i,3]
        wresponse[windex] = A[i,4]
        windex = windex + 1
# fit the regressions line and store them.
menmodel = LinearRegression()
womenmodel = LinearRegression()
combmodel = LinearRegression()
```

```
# find the parameters for the regression line
menmodel.fit(mdata, mresponse)
womenmodel.fit(wdata, wresponse)
combmodel.fit(combdata,combresponse)
# get the coefficients of determination (R-squared_
mR2 = menmodel.score(mdata,mresponse)
wR2 = womenmodel.score(wdata,wresponse)
cR2 = combmodel.score(combdata,combresponse)
print('R squared for men =',mR2)
print('R squared for women =',wR2)
print('R squared for combined =',cR2)
# variables of type LinearRegression have components called
# coef_ and intercept_ that store the coefficients and intercept of
# the model.
mcoeff = menmodel.coef_
mintercept = menmodel.intercept_
print('yhat_men = {:.4f}(age) + {:.4f}(bmi) + {:.4f}(bp) +\
{:.4f}'.format(mcoeff[0],mcoeff[1],\
                                mcoeff[2],mintercept))
wcoeff = womenmodel.coef_
wintercept = womenmodel.intercept_
print('yhat_women = {:.4f}(age) + {:.4f}(bmi) + {:.4f}(bp) +\
{:.4f}'.format(wcoeff[0],wcoeff[1],\
                                wcoeff[2],wintercept))
```

**Output**

```
R squared for men = 0.32804892866272883
R squared for women = 0.5103735991833028
R squared for combined = 0.3962220650725521
yhat_men = -0.3225(age) + 7.6564(bmi) + 1.2778(bp) +-152.1153
yhat_women = 0.7354(age) + 9.6660(bmi) + 1.6896(bp) +-306.6624

Process finished with exit code 0
```

We can see that the correlation is much stronger (a better model fit) for women than for men.

**4.**

```
# 4.

# digit recognition logistic regression
from sklearn.model_selection import train_test_split
```

```python
from sklearn.datasets import load_digits
from sklearn.linear_model import LogisticRegression
from sklearn import preprocessing
from sklearn import metrics

#get the data for all the images
alldigits = load_digits()
# get the odd rows
# we could do this with a for loop, but we can also do it with list
# splicing as below
digits = alldigits.data[::2]
targets = alldigits.target[::2]
#scales the data to help with numeric computation
data_scaled = preprocessing.scale(digits)

# Print to show there are 1797 images (8 by 8 images for a dimensionality
# of 64)
print('Image Data Shape' , digits.shape)
# Print to show there are 1797 labels (integers from 0-9)
print('Label Data Shape', targets.shape)

x_train, x_test, y_train, y_test = train_test_split(data_scaled,\
                                    targets, test_size=0.25, random_state=0)

# all parameters not specified are set to their defaults
halfdigitLR = LogisticRegression(max_iter=100)

#do the logisitic regression
halfdigitLR.fit(x_train, y_train)
# Use score method to get accuracy of model
score = halfdigitLR.score(x_test, y_test)
print('Percent correct =',score)
predictions = halfdigitLR.predict(x_test)
print('Number of predictions =',len(predictions))
cm = metrics.confusion_matrix(y_test, predictions)
print(cm)
```

**Output**
```
Image Data Shape (899, 64)
Label Data Shape (899,)
Percent correct = 0.96
Number of predictions = 225
```

```
[[19  0  0  0  0  0  0  0  0  0]
 [ 0 23  0  0  0  0  0  0  0  0]
 [ 0  0 20  0  0  0  0  1  0  0]
 [ 0  0  0 21  0  0  0  0  0  0]
 [ 0  0  0  0 22  0  0  0  1  0]
 [ 0  0  0  0  0 24  0  0  0  1]
 [ 0  0  0  0  0  0 22  0  0  0]
 [ 0  0  0  0  0  0  0 23  0  2]
 [ 0  3  0  0  1  0  0  0 18  0]
 [ 0  0  0  0  0  0  0  0  0 24]]

Process finished with exit code 0
```

It appears that we lost about a percentage point of accuracy using only half the image data.

**5.**
```
# 5.

import numpy as np
from sklearn.linear_model import LogisticRegression
from sklearn import metrics

# load the data from diabetes.txt
A = np.genfromtxt('diabetes.txt',dtype=float,usecols=(0,1,2,3),\
                  skip_header=True)
# load the disease progression into Y
Y = np.genfromtxt('diabetes.txt',dtype=float,usecols=(10),\
                  skip_header=True)
# code the Y values
n = len(Y)
for i in range(n):
    if Y[i] < 50:
        Y[i] = 0
    elif 50<=Y[i]<100:
        Y[i] = 1
    elif 100 <= Y[i] < 150:
        Y[i] = 2
    elif 150 <= Y[i] < 200:
        Y[i] = 3
    elif Y[i]>=200:
        Y[i] = 4
```

```
# all parameters not specified are set to their defaults
diabetesLR = LogisticRegression(max_iter=3000)

#do the logisitic regression
diabetesLR.fit(A,Y)
# Use score method to get accuracy of model
score = diabetesLR.score(A, Y)
print('Percent correct =',score)
predictions = diabetesLR.predict(A)
print('Number of predictions =',len(predictions))
cm = metrics.confusion_matrix(Y, predictions)
print(cm)
```

**Output**
```
Percent correct = 0.4751131221719457
Number of predictions = 442
[[  0  17   0   0   3]
 [  0 101   4   3  19]
 [  0  47   6   2  36]
 [  0  33   3   4  37]
 [  0  23   3   2  99]]

Process finished with exit code 0
```

The logistic regression does not perform as well as the linear regression for this data. This could be because the response (disease progression) is a continuous variable so that it is difficult to categorize into arbitrary ranges. It may be that the classes could be refined to improve upon the logistic model.

**6.**
```
# 6.

# neural network
print('importing packages')
import numpy as np
from sklearn.datasets import load_digits
from sklearn.model_selection import train_test_split
import matplotlib.pyplot as plt
from sklearn.preprocessing import StandardScaler
from sklearn import metrics
from sklearn.neural_network import MLPClassifier
print('packages imported')
```

```python
print('reading data')
digits = load_digits()
# get the odd rows
# we could do this with a for loop, but we can also do it with list
# splicing as below
images = digits.data
targets = digits.target
# Print to show there are 1797 images (8 by 8 images for a dimensionality
# of 64)
print('Image Data Shape' , images.shape)
# Print to show there are 1797 labels (integers from 0-9)
print('Label Data Shape', targets.shape)
print('Splitting Data')
x_train, x_test, y_train, y_test = train_test_split(images,\
                                targets, test_size=0.25, random_state=0)


# scale the data
print('Scaling Data')
scaler = StandardScaler()
# Fit only to the training data
scaler.fit(x_train)
# Now apply the transformations to the data:
x_train = scaler.transform(x_train)
x_test = scaler.transform(x_test)
# fit the network
print('fit to neural net')
digitsnetwork = MLPClassifier(hidden_layer_sizes=(60,30),max_iter=1000)
digitsnetwork.fit(x_train,y_train)
# predict and score
predictions = digitsnetwork.predict(x_test)
proport_correct = digitsnetwork.score(x_test, y_test)
print('proportion of correct predictions',proport_correct)
# get the confusion matrix
cm = metrics.confusion_matrix(y_test, predictions)
print('Confusion Matrix:')
print(cm)
# classification report
print(metrics.classification_report(y_test, predictions))
# pretty confusion matrix
rowsums = np.sum(cm,0)
scaledcm = cm/rowsums
plt.imshow(scaledcm,cmap ='Blues',alpha=0.75)
```

```
plt.xticks(np.arange(0,10,1),['0','1','2','3','4','5','6','7','8','9'])
plt.yticks(np.arange(0,10,1),['0','1','2','3','4','5','6','7','8','9'])
for i in range(10):
    for j in range(10):
        plt.text(i-.1,j+.05,str(cm[i,j]))
plt.show()
```

**Output**
```
importing packages
packages imported
reading data
Image Data Shape (1797, 64)
Label Data Shape (1797,)
Splitting Data
Scaling Data
fit to neural net
proportion of correct predictions 0.9755555555555555
Confusion Matrix:
[[37  0  0  0  0  0  0  0  0  0]
 [ 0 42  0  0  0  0  0  0  1  0]
 [ 0  0 44  0  0  0  0  0  0  0]
 [ 0  0  1 43  0  0  0  0  1  0]
 [ 0  0  0  0 37  0  0  1  0  0]
 [ 0  0  0  0  0 45  0  0  1  2]
 [ 0  1  0  0  0  0 51  0  0  0]
 [ 0  0  0  0  0  0  0 48  0  0]
 [ 0  1  0  0  0  0  0  0 47  0]
 [ 0  0  0  0  0  1  0  1  0 45]]
```

| | precision | recall | f1-score | support |
|---|---|---|---|---|
| 0 | 1.00 | 1.00 | 1.00 | 37 |
| 1 | 0.95 | 0.98 | 0.97 | 43 |
| 2 | 0.98 | 1.00 | 0.99 | 44 |
| 3 | 1.00 | 0.96 | 0.98 | 45 |
| 4 | 1.00 | 0.97 | 0.99 | 38 |
| 5 | 0.98 | 0.94 | 0.96 | 48 |
| 6 | 1.00 | 0.98 | 0.99 | 52 |
| 7 | 0.96 | 1.00 | 0.98 | 48 |
| 8 | 0.94 | 0.98 | 0.96 | 48 |
| 9 | 0.96 | 0.96 | 0.96 | 47 |
| | | | | |
| accuracy | | | 0.98 | 450 |

```
     macro avg        0.98      0.98      0.98        450
  weighted avg        0.98      0.98      0.98        450


Process finished with exit code 0
```

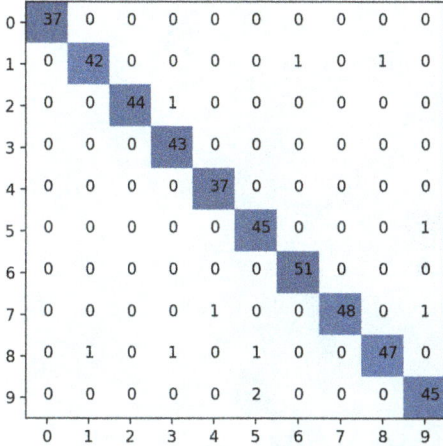

The results will vary slightly from run to run because of the random draw of testing and training sets. In this instance, the neural network was slightly better than the logistic regression.

7.
# 7.

```
# neural network
print('importing packages')
import numpy as np
from sklearn.model_selection import train_test_split
import matplotlib.pyplot as plt
from sklearn.preprocessing import StandardScaler
from sklearn import metrics
from sklearn.neural_network import MLPClassifier
print('packages imported')
print('reading data')
A_train = np.genfromtxt('mitbih_train.csv',dtype=float,delimiter=',',\
                        skip_header=False)
m,n = np.shape(A_train)
print('Training Set is {} by {}'.format(m,n))
A_test = np.genfromtxt('mitbih_test.csv',dtype=float,delimiter=',',\
```

```
                      skip_header=False)
q,r = np.shape(A_test)
print('Testing Set is {} by {}'.format(q,r))
x_train = A_train[:,0:187]
y_train = A_train[:,187]
x_test = A_test[:,0:187]
y_test = A_test[:,187]

# scale the data
print('Scaling Data')
scaler = StandardScaler()
# Fit only to the training data
scaler.fit(x_train)
# Now apply the transformations to the data:
x_train = scaler.transform(x_train)
x_test = scaler.transform(x_test)
# fit the network
print('fit to neural net')
ECGnetwork = MLPClassifier(hidden_layer_sizes=(75,50),max_iter=1000)
ECGnetwork.fit(x_train,y_train)
# predict and score
predictions = ECGnetwork.predict(x_test)
proport_correct = ECGnetwork.score(x_test, y_test)
print('proportion of correct predictions',proport_correct)
# get the confusion matrix
cm = metrics.confusion_matrix(y_test, predictions)
print('Confusion Matrix:')
print(cm)
# classification report
#print(metrics.classification_report(x_test, predictions))

# pretty confusion matrix
rowsums = np.sum(cm,0)
scaledcm = cm/rowsums
plt.imshow(scaledcm,cmap ='Blues',alpha=0.75)
k,q = np.shape(cm)
ticklist = ['0']
for i in range(k-1):
    ticklist.append(str(i+1))
plt.xticks(np.arange(0,k,1),ticklist)
plt.yticks(np.arange(0,k,1),ticklist)
for i in range(k):
```

```
    for j in range(k):
        plt.text(i-.1,j+.05,str(cm[i,j]))
plt.show()

#plot a sample of each type of ECG
imagecount = 0
reccount = 0
classnum = -1
t = np.linspace(0,186,187)
leglist = []
while imagecount < k:
    if y_train[reccount] != classnum:
        classnum = y_train[reccount]
        y = A_train[reccount,0:187]
        plt.plot(t,y)
        leglist.append('ECG Type: '+str(y_train[reccount]))
        imagecount = imagecount + 1
    reccount = reccount + 1
plt.grid()
plt.legend(leglist)
plt.show()
```

**Output**
```
importing packages
packages imported
reading data
Training Set is 87554 by 188
Testing Set is 21892 by 188
Scaling Data
fit to neural net
proportion of correct predictions 0.9767038187465741
Confusion Matrix:
[[17922   109    52    16    19]
 [  123   423     9     0     1]
 [   65     8  1351    18     6]
 [   21     0    18   123     0]
 [   31     5     7     2  1563]]

Process finished with exit code 0
```

Results will vary slightly from run to run based on the convergence of the network. Sample images for the ECG's are produced at the end of the code list.

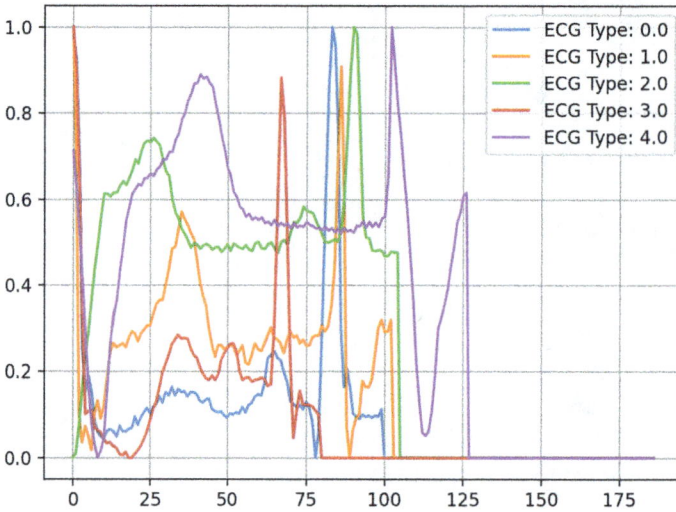

# Index

*p*-value 141
.copy 43
.format 9

accuracy 202
activation function 194
addition 3
algorithm 1
alternative hypothesis 142
analysis of variance 152
augmented matrix 59

back propagation 196
back substitution 58
bias 193

central limit theorem 125
classification report 200
coefficient of determination 170
column vector 49
comment 8
component-wise division 47
component-wise multiplication 47
concatenation 15
confidence interval 132
confusion matrix 183
correlation coefficient 170
critical region 145
critical value 144
cumulative distribution function 142

data cleaning 104
deep learning 193
degrees of freedom 138
descriptive statistics 118
differential equation 80
differential equation, order 80
differential equation, solution 80
discrete 179
division 3
dot product 49

explanatory variable 161

f1 202
false negative 202
false positive 201

floating point variable 13
for loop 85

Gaussian elimination 59
gradient descent 196

hard copy 43
hidden layer 193
hypothesis test 140

identity 52
immutable 42
implicit DE solver 89
implicit method 89
inference 118
inferential statistics 132
initial condition 84
integer variable 13
interquartile range 119
inverse of a matrix 53
iteration 71, 76

linear equation 57
linear interpolation 99
linear system 57
linear system of equations 57
local scope 17

mathematical model 161
matrix addition 44
matrix multiplication 49
matrix multiplication, component-wise 46
matrix subtraction 44
maximum 119
mean 118
mean square due to error 154
mean square due to treatment 154
median 118
minimum 119
model 161
MSE 154
MST 154
multiple regression 172
multiplication 3

neural network 193
node 193

https://doi.org/10.1515/9783110776645-010

null hypothesis  141
numpy list  25

one-tailed test  145
output to screen  8

package  11
parameter  118
partial derivative  166
phase portrait  94
pooled standard deviation  147
powers  8
precision  201
probability  123
probability density function  123
probability distribution  123
program  4

quartile  119

range  119
recursion  84
relative histogram  122
residual  164
response variable  161
Riemann sum  127
right Riemann sum  127
robust  152
roots  71
rounding entries of a matrix  55
row vector  49

sample distribution  124
scalar multiplication  48, 49
scatter plot  162

script  4
significance level  141
Simpson's rule  131
singular matrix  56
square matrix  41
SSE  154
SST  154
stability  89
standard deviation  119
standard normal  125
statistic  118
steepest descent  196
string variable  14
student T distribution  137
subtraction  3
sum of squared errors  170
sum of squares due to error  154
sum of squares due to treatment  154
sum of squares regression  170
sum of squares total  170
support  202

t test  142
testing set  188
training set  188
trapezoidal rule  130
true positive  201

variables  13
variance  119

while loop  76

zeros  71

# Index of Python Commands

.close  104

.find  16

.imshow  187

.interval  140

.predict  184

.readline()  106

.readlines  104

.round  55

.split  107

.upper  16


colon notation  15


def  17


for  85


if  29

import  11

input  19

Len  104

numpy.abs  78

numpy.array  42

numpy.dot  50

numpy.genfromtxt  119

numpy.linalg.inv(A)  55

numpy.max  121

numpy.mean  121

numpy.median  121

numpy.min  121

numpy.quantile  121

numpy.set_printoptions(precision  47

numpy.std  121


open  104


random_state  188


type  14

https://doi.org/10.1515/9783110776645-011

www.ingramcontent.com/pod-product-compliance
Lightning Source LLC
Chambersburg PA
CBHW081052220326
41598CB00038B/7063